THE PIPELINERS

THE PIPELINERS

by
FRANK MANGAN

Typography and design by the author

1977: Guynes Press
El Paso, Texas

To PAUL KAYSER
who made possible
a great achievement

BOOKS BY FRANK MANGAN

Bordertown (1964)

El Paso in Pictures (1971)

Bordertown Revisited (1973)

The Pipeliners (1977)

CONTENTS

INTRODUCTION

EL PASO NATURAL GAS COMPANY has an epic vitality that burst from a tiny regional pipeline organization into one of America's major corporations. Unusual men founded and nurtured it during a crucial period of time in the economic development of this country.

How the Company survived and flourished, sometimes in spite of almost insurmountable obstacles, is told here for the first time. The Company complied with all requests for records, withholding no facts.

Its saga is really the story of its people, their struggles and successes. Dozens of tape recordings testify to the people who built the organization and who are continuing to build it in spite of the awesome energy crisis that faces the nation today. Transcription of the tapes have been synthesized from thousands of pages to provide the inside story of El Paso Natural. The actual words of the people who built the Company are preserved here.

Behind every corporation are, after all, the people who built that organization, not the faceless shell that many Americans imagine corporations to be, but a structure of people with hearts, muscles, minds, and vision.

THOSE UNUSUAL MEN,
called wildcatters, gambled
everything in their search
for oil in unproven territory.

1

THE GREAT
PERMIAN BASIN

Wooden derricks,
iron men and red roses

THE SUN GLINTED against the two tin-roofed shacks near the weathered wood of an oil derrick, puny evidence of man's intrusion into the vast mesquite and greasewood plains of West Texas. For the two families (the Cromwells and the Locklins) who lived there in Reagan County, the dreary days dragged into tedious months, interrupted only by a cowboy or drifter who occasionally worked on the wildcat well for the payment of a meal or two.

Driller Carl Cromwell, his wife and small daughter, along with tool dresser Dee Locklin and his wife, Nora, shared the toil and the tedium, the blazing sun and the brittle, wind-swept sand in an area so remote, unproved and unpromising that experienced oilmen called it a "wildcatter's graveyard." For almost two years, the Cromwells had shared a daily spiritual and physical existence of work, eat, and sleep, as employees of the Texon Oil and Land Company. The Locklins had lived on the location less than a year, but the question naturally occurs, what motivated these families? What kept them there? One could only guess that wherever they had come from had to have been worse than where they were—or wherever they might be going.

And then it happened. Shortly before 6:00 a.m. on May 28, 1923, as Louella

Cromwell prepared breakfast for her husband it started, slowly, almost as if the earth were muttering in resentment against the bite of the drill. The roar deepened and intensified from the drilling rig 150 yards away. The two families rushed outside as a column of black smoke and a plume of oil and mud spewed wildly over the derrick.

Thus did Santa Rita No. 1 come into being; the discovery well that paved the way for the great oil bonanza in West Texas that would soon follow.

Cromwell and Locklin elatedly sealed off the bottom of the derrick with wood planks to keep out intruders and then quickly set off to lease all the available land possible from ranchers in the vicinity. Within a day and a half they acquired leases on more than 20,000 acres, and when they returned two days later everything around the gusher was black with oil which had spouted fifty to sixty feet above the top of the derrick. Their shacks, the sparse plant growth and even the Cromwell's proud flock of white leghorn chickens, and their milk cow were drenched with the heavy black gold from Santa Rita No. 1.

Dee Locklin, now retired and living with his wife in a modest white frame house in McCamey, Texas, recalls those exciting days with fond memories. "After the well came in," he said, "we felt like we were too close to the gas so we moved into town and stayed at the hotel 'til we got our houses moved a quarter of a mile farther away."

Did Lockin realize at the time what all this was eventually going to lead to? "Oh yes," he says. "I'd been through the fields but I'd never worked on one producing well. I could tell by the activity that was coming this way that it was gonna' cause quite a stir, and I was confident when I saw the Santa Rita come in it would make a good field. At first we were just tryin' to figure out what to do with the oil. There wasn't much we could do 'til we could get a tank or two out there, 'cause the oil was bein' wasted in the sand."

Word of the discovery spread swiftly and within a few days hundreds of cars inched across the Reagan County rangelands to watch the spectacle of the discovery well blowing its top.

On the 10th of June the Kansas City, Mexico and Orient Railroad ran a special excursion train ninety miles from San Angelo carrying as many as 1,000 people to the site. Box lunches and fried chicken were devoured as the spectators watched the untamed performance of Santa Rita No. 1. According to the *San Angelo Standard-Times,* "The crowd stood spellbound before the well as the oil kept coming out of the hole with such a roar that it made talking difficult unless carried on in a loud voice. Everyone seemed to have the same question on his lips: How could a man come way out here in Reagan County and stick a drill in the ground and just happen to find oil?"

Of course, it wasn't all that easy. The whole project had been fraught with serious problems from the very start.

The entire population of Reagan County

in 1920 consisted of precisely 377 people, mostly ranchers eking out a living grazing cattle on the sparse grass of this drought-stricken country. Stiles, the county seat, was so small that it didn't even have a jail at first; law breakers were simply chained to a hitch rack.

As the only village in Reagan County, Stiles grew steadily through the coming of the railroad in 1912 and the discovery of oil in 1923. But it was doomed to become a ghost town. Originally the Orient Railway included Stiles on its right of way, but when one large rancher refused permission for the rails to cross his land, the Orient changed the route and laid track through the southern part of the county.

A small railroad station, called Big Lake (from a nearby dry lake), drew residents from Stiles in the northern part of the county following the Santa Rita discovery well. Big Lake soon became a townsite and the center of oil activities. In the 1880s cow outfits reported that the lake for which the town was named was once full of water and claimed they were able to catch catfish weighing as much as twenty pounds. By 1925, Stiles couldn't muster enough votes to retain the county seat, and it died. By contrast, the town of Big Lake became a household word in the oil fraternity.

After World War I, a young attorney in Big Lake named Rupert Ricker watched with interest the great discoveries of Texas oil made earlier north and east in Burkburnett and Ranger. Perhaps, he reasoned, the rangeland of Reagan County, mostly owned by the University of Texas, also

OIL AND MUD spew wildly over the wooden derrick of Santa Rita No. 1, the discovery well that paved the way for the oil boom in West Texas.

contained an underground bounty. After
studying the geology of the area, he applied
to the university for permits to drill but he
couldn't find anyone interested in putting
up the necessary funds, some $43,000, to
develop his prospects. The roll-top desks
of even the wildest speculators thumped
shut on him, and he had less than thirty
days before his applications would expire.
Ricker's hopes sank rapidly until he encountered two El Paso men in Fort Worth:
Frank T. Pickerell, an old Army buddy,
and Haymon Krupp, a well-to-do El Paso
merchant.

Both El Pasoans showed interest in the
project, but not enough to come up with
$43,000. They agreed to pay Ricker $2,500
for his rights, which he accepted reluctantly, since he didn't have much choice.

In New York, Krupp borrowed the
necessary funds to finance the operation
and create a new corporation, the Texon
Oil and Land Company. Krupp and
Pickerell were also convinced that oil in
great quantities lay under the scrub mesquite of Reagan County. Subsequent events
proved how right they were and how right
Ricker was.

A stock salesman induced a group of
New York Catholic women to invest in
the venture. Worried about their highly-
speculative gamble, they consulted their
priest and described the Texon Oil and
Land Company which had a lease on
436,460 acres of University of Texas lands
where an oil well was going to be drilled.
The women became concerned after learn-

CARL CROMWELL was the
driller on Santa Rita No. 1.

DEE LOCKLIN, tool dresser
on Santa Rita No. 1, is retired
and lives in McCamey, Texas.

14

TEXON OIL AND LAND COMPANY well, in the Permian Basin, blows wild in the early 1930s.

ing that millions of dollars had been invested in some areas of West Texas without a drop of oil being found. The priest advised his parishioners to invoke the help of Santa Rita, the patron saint of the impossible. They did as the priest suggested, and as Pickerell was leaving New York for Big Lake, the women handed him an envelope containing a red rose blessed by the priest in the name of Santa Rita.

"When you get back to Texas," one of them asked, "have someone climb the derrick, sprinkle the rose petals over the rig and christen the well Santa Rita." Pickerell took the envelope, smiled and gave his word.

After staking out and spudding in the well on August 17, 1921, Frank Pickerell kept his promise, climbed to the top of the derrick, sprinkled the rose petals, and christened the well *Santa Rita*. When it blew its top two years later, the Permian Basin boom was on. The rest is history. Today, on the University of Texas at Austin campus, since enriched by millions and millions of dollars, the old Santa Rita pump jack is properly enshrined for all time.

The town of Big Lake, brought to life by the discovery well, still proudly claims that "this is where it all began in the Permian Basin."

Actually, the search for oil in the United States had been going on sporadically since 1859 when America's first oil well was drilled to a depth of seventy feet in Titusville, Pennsylvania by Edwin L. Drake.

IN JUNE, 1923, the Kansas City, Mexico and Orient Railroad ran a special excursion train from San Angelo so people could watch the spectacle of Santa Rita No. 1 blowing its top.

The prospecting spread from Pennsylvania to neighboring states and by the turn of the century, limited amounts had turned up in Indiana and as far west as California. By 1890, oil fever had spread to Texas, and the famous discovery at Spindletop near Beaumont on the Texas Gulf Coast in 1901, ushered in a whole new era. Spindletop became the greatest gusher of all time, and along with other wells on the coastal plain, put Texas far ahead of other oil producing states.

Those unusual men, called wildcatters, who gambled everything in their search for liquid gold in unproven territory, began sinking holes in other parts of the state, always with the hope of discovering a major field. These born gamblers moved farther and farther westward with some success, finally to the remote vastness of the Permian Basin of West Texas, where the big payoff always seemed to be just one more hole away. It would be an understatement to say that their profession was chancy, but

in Mitchell County, in 1919, they discovered the first oil in commercial quantities that far west in the state.

New towns mushroomed overnight in the huge confines of the Permian Basin. These were the boom towns with false-fronted, corrugated iron roofed buildings, with main streets that, during the rainy season, were hub-deep to the wheels of hundreds of Model T's and Dodge touring cars of the oil scouts and landmen, promoters and businessmen who were quick to get to the spots where the action was. During the spring windstorms, the caliche dust was so thick it was almost impossible to see across the streets.

A number of West Texas communities (such as Kermit, Monahans, Andrews and Pecos) were already well-established long before the boom days, but the railroad built Midland in 1881. The Texas and Pacific Railway Company rumbled west out from Fort Worth heading to El Paso in 1880. By early 1881 Chinese and Irish construction

16

LAND SPECULATORS thronged to West Texas as the oil boom moved
into high gear in 1925 and 1926.

workers appeared on the West Texas plains with their grading and rail laying equipment. A railroad mail car, serving as a post office for the construction camp, became the first official building in what was then known as Midway (now Midland)—halfway between Fort Worth and El Paso.

Odessa began when an early homesteader drilled a water well on the flat, treeless plains some twenty miles west of Midland. It became an oasis and a regular cattle stop for trail drovers and their herds of thirsty, bawling longhorn steers. The story goes that when the Texas and Pacific Railway construction crews arrived in the summer of 1881, some of the track laborers were from Russia and when they saw the wide, flat prairies it reminded them of the plains of their Odessa homeland in Russia. Out of nostalgia, they called the place Odessa and the name stuck.

In Pecos County, Fort Stockton started many years before the discovery of oil in West Texas. Nearby Comanche Spring was an Indian camp site and an oasis for white men in the Trans-Pecos region. The San Antonio-San Diego Mail Route maintained a rest stop at the spring. In 1859, the Army established and named its frontier outpost Fort Stockton. As early as 1877, homesteaders grew crops in an early attempt to use irrigated farming. When Pecos County's great Yates oil field was discovered in 1926, Fort Stockton came into its own as an oil town.

The city of San Angelo catered to the pioneer cattle and sheep raisers of the Edwards Plateau country long before the West Texas oil boom. As the wildcatters moved deeper into the western part of the state, the town shifted to become operations center for the major oil companies and independents. Later, as new prolific fields opened north and west of San Angelo, Midland and Odessa became important in their own right, and their locations in the center of the basin attracted oil company headquarters, oil field suppliers and thousands of workers and their families. In time, Midland and Odessa replaced San Angelo as the center of the oil patch.

The new boom towns were something else—towns such as McCamey, Crane and Wink. Clarence Pope, in his book *An Oil Scout in the Permian Basin,* says that two years after the Santa Rita No. 1 well came in, oil was discovered in Upton County about fifty miles west of Big Lake. This was the Johnson-McCamey well, a little north of the present town of McCamey (named after George B. McCamey, partner of Johnson in drilling the well). The town was staked out shortly after the well came in. "I saw the streets of the town of McCamey marked off, graded and paved in due time. No longer was this place known as the McCamey switch, a whistle stop along the south side of the Kansas City, Mexico and Orient Railroad. I witnessed the beginning of this bustling community; I saw huge piles of shinnery, cacti, small mesquite trees stacked high and burning curling blue smoke, with teams of mules grading to clear the way for McCamey, which has played an important part in that sector of the development of the Permian Basin. In Upton County, McCamey was one of the early new towns, if not the first, signifying that an important oil field had been discovered nearby here."

Bordering Upton County on the west, Crane County had attracted very few settlers during the homesteading years. Ranchers, even in normal years, were able to run only about fifteen cattle to a section, and this created extremely large spreads

such as John T. McElroy's sprawling 70,000 acres of sandhills and sparse grass. In 1920, only thirty-seven people composed the entire population of this lonesome, barren prairie. But with the oil discoveries in McCamey, just across the County line, perhaps there was hope for Crane.

And so there was.

In the fall of 1925, George M. Church and Bert Fields came to Crane County with a cable tool rig. Slowly but surely the unit raised and lowered the heavy bit, pounding away at the earth and literally beating a hole in the ground.

"We ran twelve hours a day, seven days a week," recalls W. W. Allman, who (although he didn't bring in the first well) drilled close to a hundred wells with the cable tool in and around Crane County. He says that Church and Fields "finished the first well in the spring of '26." With the completion of the Church and Fields No. 1 University, the McElroy Field opened.

"This field would still be flowing if we'd had the conservation we have now. Oh, we wasted lots of oil. In those days all the wells were drilled with cable tools, and we had the idea that when you shot 'em you had to let 'em flow and clean everything out. Oil just went into the sand and everywhere.

"There were no pipelines, so they stored it. This field at one time had the most storage of any in the world. It had *many* 80,000-barrel tanks. And wells flowed to 'em wide open. The gas went everywhere. The gas was all flared then and considered a waste product."

Another oldtimer, E. N. Beane, says,

"I've been accused of coming here before anybody else. Which I didn't. I did tell some newcomers that when I first got here the Pecos River was just an ant trail, and the sun was about the size of a dime, and there wasn't no moon and stars at all."

Several days before Christmas in 1926 a few tents and wooden shacks marked the foundation of Crane. A month later, there were 500 tents as the town began to take shape. J. A. Seward, one of the first four county commissioners, says, "My dad had a grocery store here in 1927. There was a meat market and a small cafe that seated about fifteen people. That was about all of the buildings.

"The county was officially organized in 1927 and when it came time to swear in the first county officials we couldn't find a Bible. So, we used a Sears, Roebuck catalog. Figured it was the next best thing.

"Wasn't no place to live, you know. We had a big old army tent we brought with us. I put in the first filling station here. We stretched that tent up down there and about three weeks later, a sandstorm came up one evening and tore that tent up in little pieces and scattered 'em along the mesquites and left us settin' right out in the open. Next morning after the tent blowed down I went to McCamey and bought a truckload of lumber. I got three of four carpenters down there and in twenty-four hours we moved into a three-room house.

"Didn't even have the gas pumps set then. I hauled gasoline out of McCamey in an old Model T truck. And brought it out in a five-gallon can and poured it in with a funnel into them old cars until I could get a pump set.

"I'd sell as much as 400 or 500 gallons a day drawin' it out in a five-gallon can and pouring it in cars."

E. N. Beane, who still feels great pride in Crane's progress, says, "Of course we had a lawless element here in those early years as did all the other towns in the area. There were gamblers and just about every type of people that follow that type of boom. We had our red light districts and there was a lot of trouble in that area.

"But the town gradually grew. Then came the Depression, and that set us back quite a bit. In fact everyone thought they were going to starve to death, including me. We survived."

The word "survived" is an understatement, for this was and *is* rugged country. You still see it in the brown weathered faces of men in khaki clothes and the caliche dust that collects on high-topped work shoes. In Crane County the sun reflects from hundreds of aluminum-painted storage tanks and many thousands of pumping jacks (they look like mechanical praying mantises), their heads bobbing up and down with alternating squeaks and groans pouring out their treasure — more than a billion barrels of it over the past half century.

Some 6,000 wells are now pumping in Crane County, and over 1,800 wells in the McElroy Field are still producing. The citizens of Crane are keenly aware of what oil and gas have done for the city. Taxes from the petroleum industry have built fine schools, and the payroll provides one of the highest per capita incomes in Texas.

"Everyone is real proud of the way things turned out here," says Beane. "We have a nice little city, nice schools, churches, a courthouse and many civic clubs. Finally we got to where we think we're civilized."

During the decade of the 1920s the search for oil marched farther westward in the Permian Basin. In 1925, the Chalk Field was discovered in Howard County, followed about a year later by the Hendrick Field in Winkler County. In 1926, wild-catters discovered the great Yates Field in Pecos County, with its first well striking oil at the unbelievably shallow depth of 997 feet. This started the Yates Field on its way to becoming one of the greatest fields on the North American continent.

The Yates discovery came on the sprawling ranch of Ira and Ann Yates, pioneers of the West Texas ranch country. Spawned by the sudden wealth from the field, a new town sprang up about a mile from the Yates headquarters. Since the new settlement had no name, a contest was held to give it an appropriate name — the winner to receive a town lot. Minnie Hardgrave won the prize, naming the town Iraan, in honor of Ira and Ann Yates. And it thrives to this day, located on the Pecos River in eastern Pecos County.

In Winkler County, living conditions for field workers were rugged — to say the least. According to Samuel D. Myres in his

THE PERMIAN BASIN of West Texas and southeastern New Mexico covers an area larger than the state of Idaho. The dark portion of this map shows the general outline of the Basin and includes within it the Delaware Basin, Midland Basin, Val Verde Basin and other oil and gas producing areas.

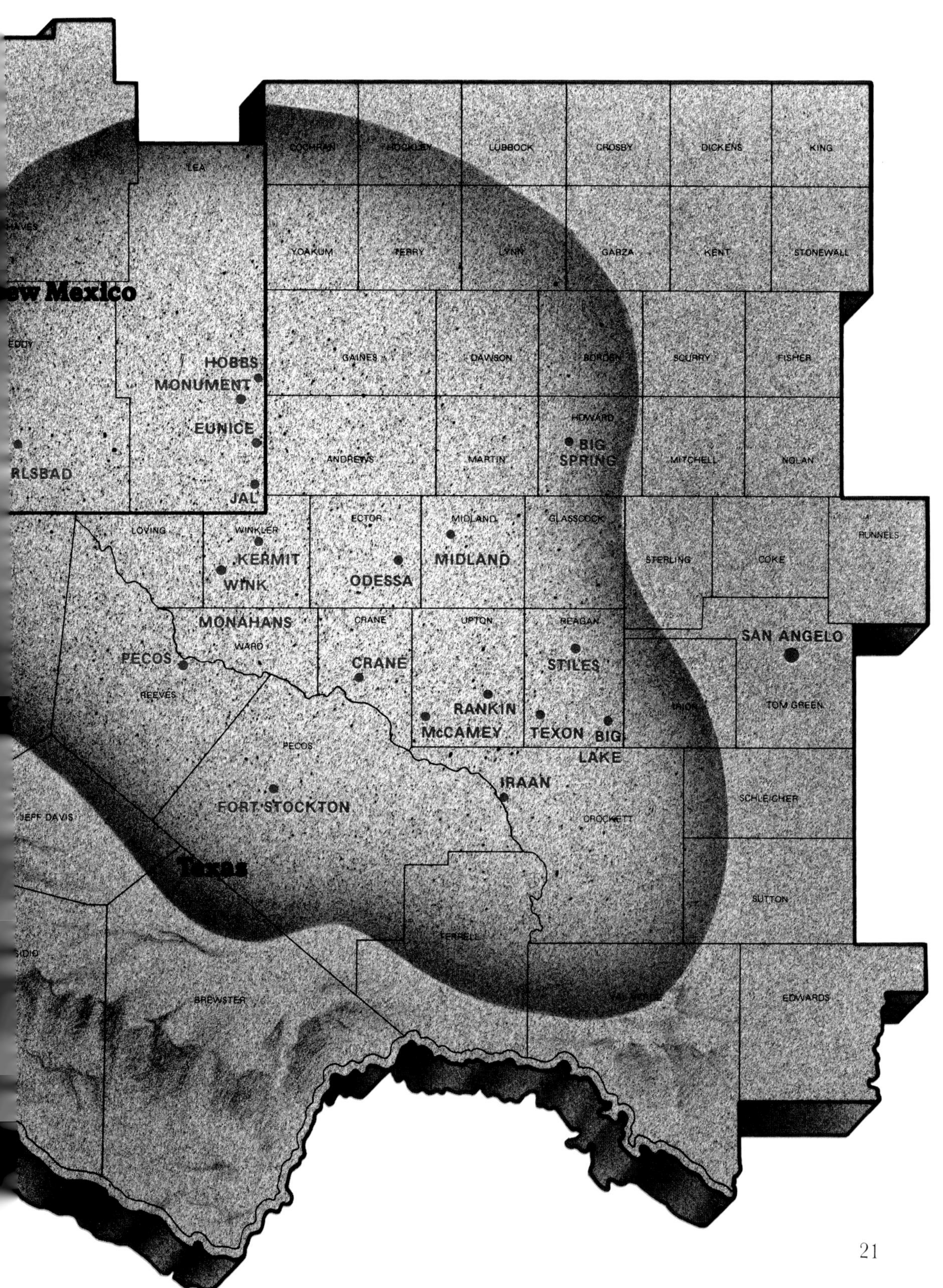

New Mexico
Texas
LEA
HOBBS
MONUMENT
EUNICE
JAL
CARLSBAD
EDDY
HAVES
WINKLER
KERMIT
WINK
LOVING
MONAHANS
WARD
PECOS
REEVES
PECOS
FORT STOCKTON
JEFF DAVIS
BREWSTER
PRESIDIO
COCHRAN
HOCKLEY
LUBBOCK
CROSBY
DICKENS
KING
YOAKUM
TERRY
LYNN
GARZA
KENT
STONEWALL
GAINES
DAWSON
BORDEN
SCURRY
FISHER
ANDREWS
MARTIN
HOWARD
BIG SPRING
MITCHELL
NOLAN
ECTOR
MIDLAND
GLASSCOCK
ODESSA
MIDLAND
RUNNELS
STERLING
COKE
CRANE
UPTON
REAGAN
SAN ANGELO
CRANE
STILES
TOM GREEN
RANKIN
McCAMEY
TEXON
BIG LAKE
IRVION
IRAAN
CROCKETT
SCHLEICHER
TERRELL
SUTTON
EDWARDS

DURING THE LATE 1920s, Wink, Texas, enjoyed the dubious reputation of being one of the wildest

book, *The Permian Basin,* tons of heavy equipment were hauled in by wagons and mules, or trucks, generally in clouds of dust. While the major companies such as Magnolia (now Mobil), Humble (now Exxon) and Gulf maintained large camps near the drilling and production sites for their employees, says Myres, "the largest number of workers by far lived in or near the town of Wink, which was located on the southwestern edge of the Hendrick Field. Although Wink perhaps deserves the accolade of the toughest place in the Permian Basin, it was by no means all bad.

Myres continues, "The town began to develop shortly after the Gulf gusher started flowing late in March, 1927 . . . A gang of workers was busy grading the streets, as well as the roads leading out of town. It grew at a rapid pace until in 1928 its population increased to an estimated 15,000, making the town one of the largest, and most dynamic of the period.

"At the beginning, residential lots sold for $50 each, more or less, but during the boom the prices rose to approximately $150 per lot. Commercial buildings were usually constructed of galvanized iron, with false fronts. Dwellings, including tents, which were numerous and scattered all around, and shotgun houses with sheetrock walls, unsealed on the outside and tin roofs — plus a few substantial wooden structures. The buildings that contained large amounts of tin or sheet iron became unbearably hot in the summer. Some relief could be had from electric fans, but air conditioning as it exists today was unknown. Sleeping out-of-doors was a common practice during the summer months."

In addition to the oil field workers and their families and the town businessmen, scores of bootleggers, gamblers, prostitutes and racketeers plied their trades openly — often with the sanction of corrupt local officials. During these years Wink enjoyed the dubious reputation of being one of the wildest settlements in the Southwest, a boom town whose breath reeked with bootleg whiskey and sulphurous gas. Located in the only masonry and fireproof building in town was the Rig Theater, which showed the latest silent pictures to large audiences. The downtown streets

settlements in the Southwest, a boom town whose breath reeked with bootleg whiskey and sulphurous gas.

were unusually crowded, even at midnight when the twelve-hour shifts changed in the nearby fields. According to Sam Myers, "Many townspeople parked their cars along the streets, remaining inside the cars to see the mass of humanity pass by. Women frequently came to the downtown area alone, even at night. Although many rough characters frequented the area, women were usually safe. The frontier tradition of respect for females prevailed throughout the boom period."

But the boom in Wink was short-lived. In 1930, the Depression, coupled with the opening of the East Texas oil fields and the movement of the Permian Basin oil play northwestward to the Keystone Field, caused a mass exodus from the town. The dust settled back on the board sidewalks. In short, the town of Wink dried up. Literally, Wink even holds the all-time record in the state of Texas for dryness; in 1956, the total rainfall for the year was only 1.76 inches. Today about 1,000 persons remain, but the spirit of past glory still lingers in their memories.

While the prolific fields around Wink poured forth their wealth in the late 1920s, one major area remained to be opened in the Permian Basin. This was Lea County, New Mexico, just across the state line from Winkler County. The Maljamar Oil and Gas Corporation made the first discoveries. In the extreme southeastern corner of New Mexico, the Texas Company opened a new producing area with the drilling of its No. 1 Rhodes well, six miles southeast of the tiny village of Jal. The Rhodes well was completed for 22.5 million cubic feet of gas and 300 barrels of oil daily. The Texas Company completed a large gas well, the Eaves No. 1, in June, 1928, in the same general area — good for ninety million cubic feet daily. Immediately north of Jal is the Eunice area. Near Eunice, Continental Oil Company successfully found both oil and gas, the largest of these wells producing seventy-seven million cubic feet of gas daily.

The little town of Eunice grew and prospered with the increasing operations in nearby oil and gas fields and attracted a number of processing plants of the major oil companies.

DIRT STREETS and false-fronted buildings (left) were typical of downtown McCamey in 1927. The structure in right foreground with "bed" sign is a cot house which slept oil field workers three shifts a day. Above, the town of Iraan is still an oil and gas center a half century after the discovery of oil on the sprawling ranch of Ira and Ann Yates. Below, Santa Rita No. 1 pump jack is properly enshrined for all time on the University of Texas campus in Austin.

Eighteen miles north of Eunice and almost straddling the Texas state line the little settlement of Hobbs was soon slated to profit by and grow with the Permian Basin boom. The boundary between Texas and New Mexico was officially surveyed in 1911, and Lea County was created in 1917 from parts of Chavez and Eddy counties. In 1927, the first petroleum survey crews to arrive in what is now the city of Hobbs found only one house, two trees and one windmill. The place originated in 1906 when James Hobbs and his family acquired homestead rights in the area. They arrived in covered wagons from Brown County in central Texas and were among the first settlers in what was then eastern Eddy County.

The coming of the oil men gave Hobbs special flavor and along with the usual influx of riffraff, drifters and fast-buck artists, made the new town uncommonly noisy and rowdy. Hobbs quickly exploded into the last great boom town in the Permian Basin. The first established improvements in the city of Hobbs occurred when James Hobbs built a store at the corner of what is now Marland and Dal Paso streets. The first post office in the community was in his store.

Unlike many of the Permian Basin boom and bust towns, Hobbs grew and it prospers to this day. The first passenger train of the Texas and New Mexico Railway arrived in 1930 and opened up to a new era this previously remote and isolated town, one of the few remaining regions of the old Western frontier. Today, 30,000 Hobbs residents live in a modern progressive city with broad paved streets, beautiful homes, tree-shaded parks and modern shopping centers — visual proof of its continuing vital importance in the oil and gas business of Lea County, New Mexico. The Hobbs Field still yields large quantities of oil and gas some fifty years after production began.

As the decade of the 1920s drew to a close, it became increasingly evident that literally billions of cubic feet of natural gas were being wasted in the process of producing oil. Gas was still regarded as a nuisance to be disposed of as rapidly as possible. What the operators wanted was oil for which there was a market where the big money was.

Most wells in the Permian Basin produced gas in conjunction with oil. The gas escaped in much the same way gas pops out of a bottle of carbonated beverage when the cap is removed. Since there was no system of storing, collecting or marketing the natural gas, oil operators separated it at the wellhead and burned it in thousands of huge flares that lighted up the whole basin and provided an awesome glow at night for many years. In the Kendrick Field alone oil operators wasted 200 million cubic feet of gas daily simply because there was just nothing else to do with it.

Later, this loss of one of America's irreplaceable and most valuable natural resources provided a major supply of gas for El Paso Natural Gas Company, destined to become one of the country's largest corporations.

DANCE HALL sits in splendid isolation
in southeastern New Mexico.

SANTA RITA No. 1 is still
a landmark in Reagan County.

CHURCHGOERS from miles around
attend services at Orla, Texas on the
western edge of the Permian Basin.

REMAINS of early-day motor courts can
be seen in Hobbs, New Mexico.

COLORFUL PAINT JOB entices
customers at Jal Recreation Center.

A REMINDER of the past, this truck reposes
among weeds in far West Texas oil field.

DESERTED STONE courthouse
is one of few buildings remaining in
Stiles, Texas.

JAL PICTURE SHOW in 1973.

THE OLD AND THE NEW in West
Texas' Oil Patch.

THE PERMIAN BASIN

An underground look

Vast is the word for the Permian Basin. In West Texas and southeastern New Mexico alone, it covers more than 88,000 square miles, an area larger than the state of Idaho. For a general concept of its limits, it extends from Sweetwater, Texas, on the east, to Artesia, New Mexico, on the west, and from the High Plains of the Texas Panhandle on the north, to Brewster County on the south.

What is a basin? A basin is a depressed area that may lie at ground level or below ground.

Geologists recognize three types: One is a topographic basin, a depressed area on the surface of the earth. The others are structural and depositional basins. The latter two are underground depressions in which the formations are slanted from all sides toward a central low point. The Permian Basin combines depositional and structural characteristics; both have the depressed appearance, but their formation was caused by different forces.

Rocks of Permian age are found in many parts of the world, but the Permian System was identified more than a century ago in the province of Perm in Russia. The province gave its name to the Permian System, but the Russians have since renamed the system Molotov. Studies of Permian sediments in America's Southwest began about forty-five years ago when geologists first examined the limestone and other sedimentary rocks of Permian age that outcrop at El Capitan, a huge bluff about 100 miles east of El Paso. Here, the western edge of the basin formations dip or slant underground toward the east. When geologists examined the east side of the basin, they found that the same formations incline downward, toward the west.

Some 250 million years ago during the Permian Period, a huge inland sea blanketed much of what is now Texas and southeastern New Mexico. Rising above the water during early Permian time was a giant thumblike ridge 150 miles long and forty or fifty miles wide, that divided the Permian sea into two smaller basins—now known as the Midland Basin and the Delaware Basin. The ridge itself, which geologists call the Central Basin Platform, is today a subsurface feature underlying an area in southeastern New Mexico and West Texas extend-

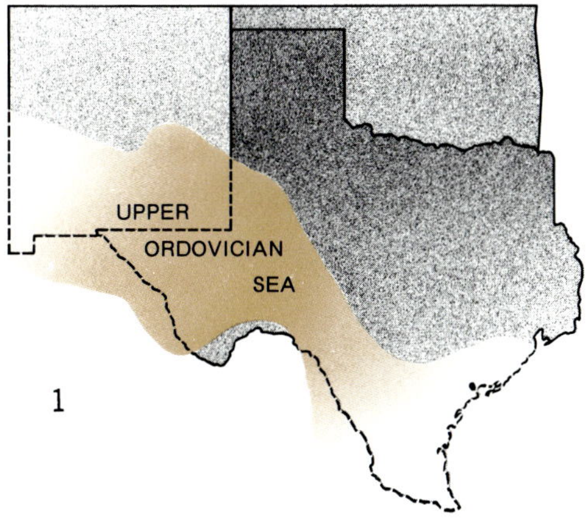

THESE MAPS show generalized locations of some of the major seaways that covered what is now the Southwestern part of the United States in the distant past.
1. The Upper Ordovician Sea about 445 million years ago. 2. The Late Pennsylvanian Sea about 275 million years ago. 3. The Permian Sea about 230 million years ago.

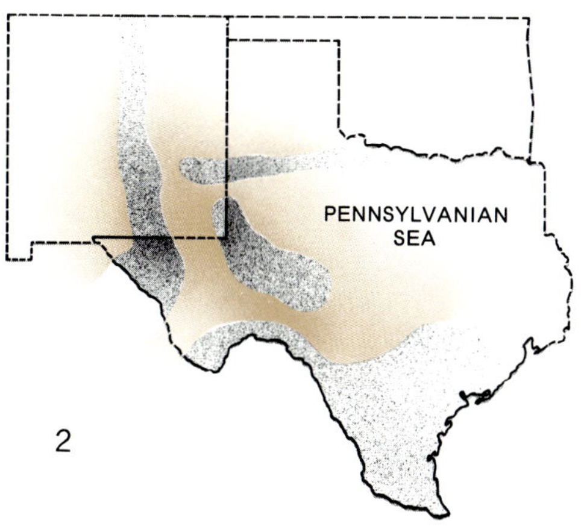

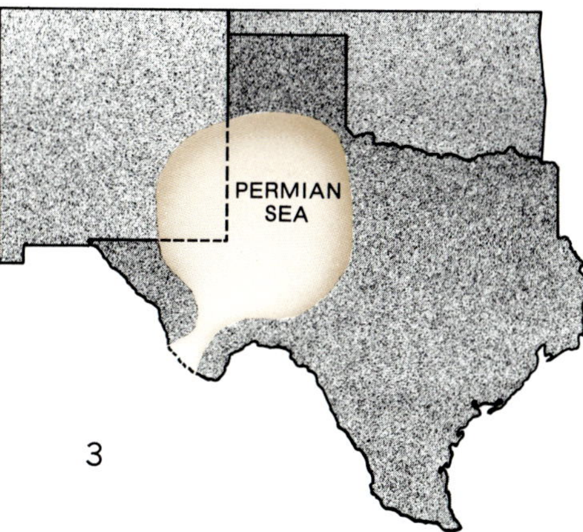

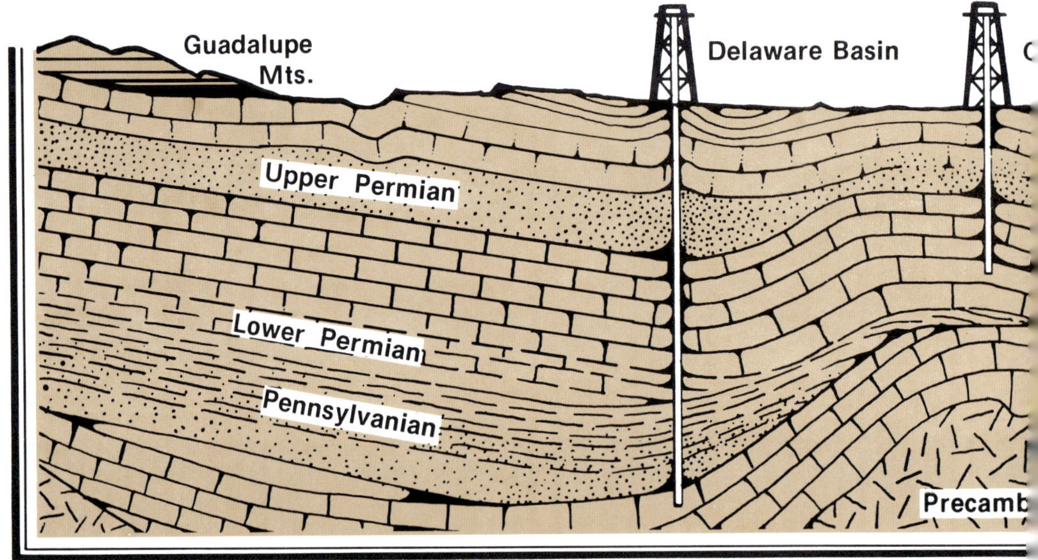

THIS EAST TO WEST diagrammatic cross section shov

ing from Hobbs, New Mexico, on the north, south through Jal and Monahans, and on southeast to near Iraan, Texas.

Had man been present during very early Permian time, he would have found the Odessa area situated along the shoreline of a magnificent body of blue water extending eastward further than his eyes could see. What is now a semi-arid expanse of scrub brush and ranches was once a paradise of mountain ranges covered with trees. The surrounding shallow seas were studded with reefs teeming with marine and plant life.

On the western edge of the Permian sea grew a great barrier reef, part of which has since been uplifted and stands today as a magnificent peak piercing the sky several thousand feet above the old Permian sea level. This reef, the Capitan Reef, is prominently exposed today in Signal Peak which rises immediately in front of Guadalupe Peak, the highest point in Texas. During Permian time, it was at, or slightly below sea level, like the present day Great Barrier Reef of Australia. To the north is what is known as the Northwestern Shelf. It was similar to the Central Basin Platform, an elevated area later submerged under a shallow part of the sea. The same situation existed on the eastern side of the Permian Basin in what is now known as the Eastern Shelf. Originally, that area was also elevated but it, too, was eventually covered by a part of the sea.

As millions of years passed the sea gradually moved to the north and southwest and covered not only parts of Texas and eastern New Mexico, but advanced

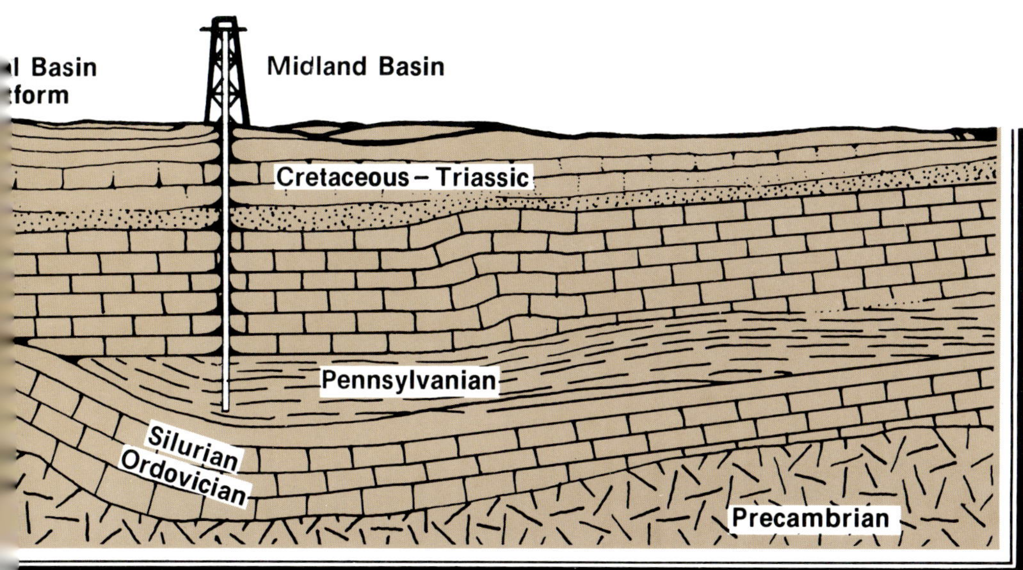

eralized view of the nature of underground formation in the Permian Basin.

into western Oklahoma, central Kansas and into Nebraska. During this time the Central Basin Platform, which once extended above the sea, was submerged under water and the entire Permian Basin was receiving deposits of sediments and organic matter which sifted down into the sea bottom and ultimately resulted in the great accumulations of oil and gas which have been providing a big part of America's energy for more than half a century.

The oil and gas that pours forth from the Permian Basin today formed slowly through millions of years in the sediments that were deposited in this great ocean. Tiny plants and animals lived in the shallow waters of the sandy beaches and along rugged seacoasts, just as they do in today's oceans. As these plants and animals died, their remains settled, to decay, on the muddy bottom. Fine sand and mud washed down from dry land areas to cover these organic remainders. As the sediments thickened, their great weight pressed them into hard, compact layers. Bacteria, heat and pressure gradually changed the entombed plant and animal matter into oil and gas. Through millions of years, other sediments formed over these shales (source beds) sealing off the oil and gas within adjacent reservoir rocks. As the prehistoric seas drained away, dry land appeared to cover the petroleum deposits.

Toward the close of the Permian Period, when the sea became more shallow, huge quantities of sea water evaporated, causing various kinds of sediments to be deposited on the sea bottom, ultimately forming thick beds of gypsum, salt,

ANCIENT sea life can be seen in this fossiliferous limestone which was deposited on the shelf areas of the Pennsylvanian Sea in the Permian area. Rocks like this can be found today, exposed at the surface, in mountains ranges such as the Guadalupe, Delaware, and Glass mountains, all in close proximity to the Permian Basin.

anhydrite and potash, particularly in the last remnants of the old Permian Basin Sea near the Carlsbad, New Mexico area.

Carlsbad Caverns are also remnants of the Permian age, and they owe their existence to openings made by underground water in Permian limestone formations. Settling of the original sediments and repeated earth movements made cracks in the limestone rocks. The caverns began to form when fresh water from above ground found its way through the pores of these rocks.

By the end of Permian time, the basin was above water. However, it was still a basin, and streams washed earth down from higher land areas causing great thicknesses of non-marine sands and clays to be deposited. Drillers in the West Texas area now know these shallow formations as "red beds."

The Permian Period was not the last great geological time through which the area passed. Following the Permian, during a geological period known as the Cretaceous, another sea covered most of the state of Texas and much of the Rocky Mountain region. At this time, thick beds of sand and limestone were laid down and later seas moved in and out of the area.

Slowly, however, as millions of years passed, the Permian Basin again emerged from the sea, this time to stay. But completely buried under thousands of feet of rock were the Central Basin Platform, the Midland and Delaware basins and the Shelf area. These buried regions today are the source of oil and gas.

THE TIN LIZZIE

It stimulated the search for oil

THE FIRST STANDARD OIL
service station was set up in a
portable garage in Columbus, Ohio
in 1912. Motorists drove in the
front door, had their tanks filled and
drove out the back.

In 1893, Charles and Frank Duryea rigged a one-cylinder gasoline engine on a
$70 secondhand carriage and the automobile came to life in America as a simple
means of transportation. But even from the humble beginnings of the automobile
age there were signs of future traffic problems. In 1895, there were only four
gasoline-powered vehicles in the United States, and two of them managed to
collide in St. Louis, injuring both drivers.

Henry Ford launched the auto age in earnest in 1908 when he rolled out the
first Model T—a symbol of speed, power and luxury. It sold for just under a
thousand dollars. By 1914, Ford put his entire production on an assembly-line
basis, and by 1924 he could put a car together in his plant in ninety-three minutes.
You could buy a Model T for $265. By 1925, Ford produced 9,000 cars a day,

OLD TIME horse-drawn Texaco oil truck (above). Below, as a variant to the shout of "get a horse", this farmer with a team of oxen pulls a horseless carriage and its occupants after a breakdown in the country.

one approximately every ten seconds. Those who couldn't afford the price bought their beloved Tin Lizzies secondhand for $25 and up. Henry Ford built fifteen million of them and led the country into the automobile age.

Although the invention of the internal combustion engine and changeover from manpower to machine power in the nineteenth century touched off the great search for oil and gas, petroleum products of a sort had been around for centuries. The ancient Egyptians found petroleum seeping to the surface from underground springs, and they used the thick, gummy substance to coat mummies with the natural asphalt. The Chinese found natural gas while drilling for salt, and some 2,000 years ago they piped the gas from shallow wells through bamboo poles. They burned the gas under large pans to evaporate sea water for salt.

Centuries later, American frontiersman Kit Carson collected oil from a seepage in Wyoming and sold it to pioneers on their way West as axle grease for their covered wagons. From the days of colonization until the Civil War, most American homes lit glass-chimney whale-oil lamps and dim tallow candles for illumination. It was still an age of going to bed when the sun went down. Since there were no substitutes, America was literally at the mercy of the whaling industry. And even in those days of no computers or energy experts, it didn't take too long to figure out there just were not enough whales to go around. The result was that the price of whale and sperm oil rose from forty-three cents a gallon in 1823 to more than $2.55 a gallon in 1866.

And what happened? We were forced to seek substitutes. So, we discovered kerosene and started the petroleum period. In the 1880's, kerosene for lamps was the chief product of the petroleum industry. No practical use had yet been found for gasoline, a by-product of kerosene, and gasoline was often dumped into creeks and rivers to get rid of it.

The beginning of the natural gas industry was a different story. An early discovery of natural gas was made in Fredonia, New York, about forty miles from Buffalo. In 1821, a gunsmith named Aaron Hart drilled a well and when he reached only seventeen feet below the surface a hissing sound indicated the first natural gas well discovery on the American continent. The enterprising Hart experimented with the gas to find possible uses. Using hollowed-out logs as a crude pipeline he provided nearby buildings with light, and by the turn of the century, the soft glow of gaslights illuminated the streets and homes of much of America.

In 1900, gas and electric lights rapidly replaced kerosene lamps as the auto-mobile rolled onto the American scene. Only about 8,000 of these strange looking horseless carriages bounced precariously through the mud and deep ruts of America's dirt roads, but within ten years there were almost a half million auto-mobiles replacing the horse and buggy. As Americans toured the country they quickly discovered that the roads were not good enough. Some motorists in West

LOTS OF COURAGE was required to navigate early American roads.

Texas followed the ruts of the old Butterfield Trail, and wise drivers took their chances armed with picks, shovels and axes. High centers had to be dug out and an occasional mesquite tree had to be chopped down. And there was the ever present danger of getting stuck in the powdery, runningboard deep sand. But in 1921, a federal highway act enabled a connected cross-country highway system to accommodate the new motoring public.

America's new motorists found themselves puzzled and dismayed trying to find their way across the country, since there were few road signs and even fewer road maps. In the East various colors were painted on trees, fences and telephone poles, indicating through roads. But in much of the West there just weren't enough trees and fences to color code. In 1925, the Federal Government devised a system using shield-shaped signs to designate through routes crossing the various state lines, identifying them as U.S. Highways. State roads were designated by circular signs. Even-numbered highways went east-west and odd-numbered routes went north-south. Many of the new highways had names, such as the old "Bankhead Highway" that traversed Texas from east to west and later became U.S. 80. The roadbuilding of the 1920s became the foundation of most of the interstate highway network of today.

It was an age of transition. As the horse and buggy began to disappear, livery stables became garages, catering to both horses and automobiles. Blacksmith shops branched out into a new trade, selling gasoline. Originally gasoline and motor oil was sold in containers carried out to the curb and poured by hand into

NTLEMEN fill their own tank.

GRANDIOSE tire displays attracted motorists.

the cars. Soon curbside hand pumps appeared, and crude service stations sprouted throughout the land, most of them little more than sheds with no uniformity of design or operation. The attendant usually wore faded bib overalls. Automobile owners greased, change tires and tinkered with their own cars, since the entire inventory included one grade of gasoline, one kind of crankcase oil and one- and two-pound cans of grease.

Proprietors gradually added other niceties such as fan belts of different sizes, an outhouse for the convenience of travelers on dusty roads, a metal box containing bottles of Orange Crush floating around in water cooled down a few degrees by several chunks of ice, a Coca Cola thermometer nailed to one of the wooden canopy posts and an ever-present metal sign advertising Beechnut Chewing Tobacco. Hand-lettered messages contributed their own brand of humor with such classics as "We don't know where mom is, but we've got pop on ice."

The super service station soon made its appearance. Foreshadowing a distinctively American institution, Standard Oil opened one model in 1912 at Memphis. It boasted thirteen pumps, a ladies' restroom and a maid who served ice water.

America in 1920 was on the move, free at last to overcome the isolation in which most people had always lived. The gulf between city life and rural life narrowed. Twenty million motor vehicles hit the road by 1925 and in the frenetic search for new discoveries of oil, wildcatters zeroed in on the vast, untapped expanses of the great Permian Basin of West Texas and New Mexico.

WEST OF THE PECOS,
pipeliners found only a
hellish mixture of solid
rock mountains, deep
powdery sand, rattlesnakes
and salt flats.

2

PAUL KAYSER

The first pipeline

I N THE LATE 1920s it seemed, like the song's lyrics, that God had literally shed his grace on America, from its purple mountain majesties and from sea to shining sea. Even though part of its rural areas and large cities still had their share of poverty, the great American dream was at last coming true for millions.

The hardships and sorrows of World War I, its memory dimmed by more than a decade of progress, brought on the most spectacular economic boom the country had ever experienced. The decade began when Warren G. Harding was elected President on a platform of "back to normalcy" and in the freewheeling mood of the Twenties, Americans were no longer interested in the problems of the world.

They craved excitement and then got it any way they could. Even the dry shadow of the Volstead Act and Prohibition that settled over the land succeeded only in speeding up the hedonism of American youth. Bootleggers and private speakeasies replaced the saloons.

In Chicago, Al Capone's gang killed six rival gangsters in the 1929 "St. Valentine's Day Massacre," an act intended to strengthen Capone's control of Chicago's thriving liquor business.

F. Scott Fitzgerald chronicled the era as the "greatest, gaudiest spree in history."

Girls, in particular, rebelled and changed their passive role of morality from high-button shoes and hobble skirts to become jazz age flappers with short skirts, silk stockings and bobbed hair. They danced to the scandalous Charleston rhythms of sensuous jazz, drank bathtub booze and smoked cigarettes at all night parties.

A youthful pilot named Charles Lindbergh amazed the whole world with the first solo, nonstop airplane flight across the Atlantic in 1927. Hollywood produced the first talking picture, *The Jazz Singer*, starring Al Jolson, that same year, and by the end of the decade the silents flickered out.

The jazz age typified America at its zaniest, and for many, its most affluent. It was an age when seedy little bands played music for thousands of bone-weary marathon dancers dragging their partners like limp Raggedy Ann dolls across waxed floors from coast to coast. Fad-happy crowds, eyes glazed, watched with glee as flagpole sitters like Shipwreck Kelly perched fifty or so feet off the ground, each trying to outdo the other and set a new world record.

The automobile transformed the country's entire transportation system, as America became a nation of giddy consumers. With the mass production of goods, corporate profits zoomed, as did the life styles of working men and women. In 1928, Herbert Hoover defeated Al Smith to become President of the United States. The stock market soared to unprecedented heights, indicating that the abolition of poverty in America lay just around the corner.

RCA stock climbed from less than 100 to 400 between March and November. During the first part of 1929, prices on the New York Stock Exchange surged ever higher. Stock speculation became a mania throughout the country with thousands of small investors pouring their savings into the market. By September, unheard of heights had been reached. RCA topped 500. General Electric, which had sold for about 128 in 1928, now approached 400. Many others soared accordingly. A few conservative brokers expressed alarm that stocks were becoming far overpriced, but nobody listened. Prosperity seemed to be limitless.

Meanwhile, back in the vastness of West Texas and the Permian Basin, the oil boom continued. Drought-stricken ranches that only a few years earlier had barely provided a living for tough homesteading pioneers and their families were now pouring forth huge amounts of oil to fuel the nation's expanding economy.

None of this escaped the notice of a bright young lawyer in Houston, Texas named Paul Kayser, a man of boundless energy and confidence, destined to become one of the giants of the natural gas industry.

Some of his employees later described him as a man "five-foot tall with a ten-foot stride." Kayser's disarming smile has a habit of spreading across his strong, square jaws. Two sparkling brown eyes seem to look right at you with a preconceived knowledge of what you're going to say. Always immaculately groomed, he hardly resembles the Hollywood version of a raucous, rowdy, freewheeling Texas oil tycoon, yet his enthusiasm and basic honesty have made him one of the great men of the oil and gas industry.

In 1973, some forty-five years after the formation of El Paso Natural Gas Company, the Permian Basin Museum's Hall of Fame in Midland, Texas, inducted him into its halls. Berte R. Haigh, chairman of the awards committee, put it this way: "In 1928, Paul Kayser saw something that very few men of his age saw. In those days, natural gas was the ugly sister of the best looking girl in town. Nobody wanted her around, but she came along anyway. Kayser is a man who is a conservationist in the broader and better sense because he has helped conserve energy supplies as well as the talents and efforts of people around him, the prosperity of the Permian Basin and the future of his state and nation."

During the awards presentation before a capacity crowd, Morgan Davis, former chief executive of Humble Oil and Refining Company, recalled the early days of the oil fields. He said a young man from the East came to make his fortune in the oil fields of Texas. He wrote to his father, "The West looks like a good place to settle. However the countryside needs two things; shade and more water." His father wrote back, "That's all hell needs, son."

Men with vision, such as Paul Kayser, saw the raw land in a different light. Davis said of his good friend, "Paul Kayser is the best known and most highly respected man internationally in the gas industry."

This respect has run through the years, when Kayser made hundreds of major business deals, all of them with the same basic premise: A sound business deal must be good not only for you, but for the other party. According to Kayser, "I start out on this principle: Find out about the man you're trying to contract with — what he really wants and what is good for him. When you learn that, then aim to give it to him. But at the same time he must give you the things you want. But don't ever get the attitude of magnifying what you want and minimizing what he wants. A man who has made a bad contract is going to sit up nights trying to find a way to break it. If it is a good contract, you can make it much more rapidly, and in the end it will be a good thing for you as well."

Kayser was born in the piney woods of deep East Texas near the little town of Tyler when the West was still a frontier. He grew up on a farm nearby and became a school teacher after graduation from Baylor University. After a brief two years as principal of Gatesville, Texas High School, from 1909 through 1911, he studied law by correspondence from the University of Texas and was admitted to the Texas Bar in 1915. He was soon practicing law in Houston as a member of the firm of Huggins, Kayser and Liddell, until World War I interrupted his career and he served as a

Paul Kayser

H. G. Frost

captain in the 7th Cavalry of the National Guard. After the war, he returned to Houston and remained with the firm until 1929.

During the early 1920s, as an attorney, he helped organize and launch a new organization, Southwest Gas Utilities Company — headed by J. W. Colvin. Through his law firm, Kayser also met an unusually gifted man named H. G. Frost, who headed up the Salt Dome Oil Company. Frost had been successful in previous oil lease ventures, particularly in East Texas and Louisiana, where in spite of the pitfalls of speculation and gambling in oil land, he always seemed to land solidly on both feet. He was the kind of man who would gamble while lesser men talked.

A basic truth of the petroleum business was as valid then as it is today: This is a high risk business requiring daring and capital, but the public still doesn't understand the risks involved. It takes a special breed of men to put a million dollars into a hole and find out it's dry—they've blown it and they're back running a hamburger stand. Statistics show that only one out of ten wildcat wells drilled by exploration crews in the United States is successful. The rest of the wells are losers, just dry holes, part of the expense and disappointments of the gamble. But H. G. Frost had faith in his great ability and the future of the oil industry.

Early in 1928, Kayser and Frost made a New York trip to put together a company, a three-way partnership with Colvin, for the accumulation of royalties on gas and oil. This entailed finding enough financing to purchase leases on land that might prove profitable in drilling for oil and gas. Royalty agreements would be sold and issued to speculators willing to gamble that the effort would be successful enough to earn a substantial profit. Millions of dollars were made, sometimes almost overnight, even by hard-scrabble farmers and factory workers who purchased shares in unproven areas of the vast Permian Basin.

During this meeting in New York, Kayser and Frost planted the first seeds for what was later to become El Paso Natural Gas Company. In his own words, the peppery Paul Kayser tells it like it was: "We were in the office of the brokerage company that had financed the Southwest Gas Utilities Company. The gentleman told us that he wouldn't be able to sell the stock of a royalty company. 'But,' he said, 'I'll tell you what you *can* do. If you've got any natural gas property, or if you're able to put together a natural gas company, we will finance it for you just like that.'

"So," says Kayser, in his soft East Texas accent, "we went over to a side of the wall and looked at a map of Texas. It showed that every single city in the state had natural gas except El Paso. The broker asked us if we knew where we could get some gas.

"Frost wrinkled his brow and answered, 'Well, there is some gas out where I was buying these royalties. Out in the Permian Basin.'

"With that," said Kayser, "we decided to look into it. We told the broker we would be back if we could put something worthwhile together."

The pair climbed aboard the next train from Grand Central Station back to Texas. The sound of the wheels against steel rails hammered out a monotonous beat in their Pullman car as they sped through the night and all the next day along the East Coast and then westward to Houston. Uppermost in their minds were two restless thoughts: Could they find enough gas far out in the western edge of the Permian Basin to supply El Paso, and if so, could they do it before somebody else developed the idea?

Several days after their return to Houston, Frost came to Kayser with bad news. Another group had just been granted a franchise to sell gas to El Paso. They had the money and the contracts to build a pipeline west to the City from the Permian Basin. The announcement was met with the cheerfulness of a person whose doctor has just informed him that he needs an immediate gall bladder operation.

"It looks like we're already beaten," said Kayser, "I don't want to get into a fight, so let's just forget the whole thing."

But fate intervened. Within a month after Kayser had given up his pipeline dream, he picked up bits and pieces of information on the El Paso pipeline plan. Kayser learned that perhaps all was not lost. The latest situation was simply this: No, El Paso was not yet supplied with gas —and not only that, nobody was really in a position to build a company to supply the City.

"The next thing I did," recalls Kayser, "was to call El Paso and talk to Volney Brown of the law firm of Goggin, Hunter and Brown, representing El Paso Electric Company, because he could give me an accurate report. Brown said he didn't know the situation but he'd talk to the mayor.

"He called me back in about thirty minutes and said, 'I talked to Mayor R. E. Thomason, and he says they haven't granted a franchise to anybody, and he will be glad to see you.' "

Kayser was buoyant. "We started all over," he said, "and set out to find the gas that would support a line to El Paso. We quickly made supply contracts from three gas wells near the little village of Jal, New Mexico, just across the Texas line. That was it."

They obtained firm contracts to supply El Paso's industries with natural gas since these would consume the largest portion. Southwestern Portland Cement Company, El Paso Brick Company, International Brick Company, El Paso Electric Company, key industrial plants in El Paso, and many other smaller companies eagerly sought the product. They signed contracts to receive the new fuel and provided a base of income for the new company to meet its commitments. One holdout was the local artificial gas distributing company, Texas Cities Gas Company. They apparently didn't believe that Kayser's group had the necessary financial clout to succeed.

"But nevertheless," says Kayser, "we went ahead. We put $50,000 in the State National Bank. I remember the day I deposited the money, the bank vice president asked me, 'Are you just putting this in here for a day or so?'

"I told him, no sir, I'm just trying to open an account with you."

The next step involved a much-needed franchise from the City council of El Paso, which included A. B. Poe, R. E. Sherman, R. E. McKee and Stewart Berkshire.

"When we appeared before them," Kayser recalls, "we presented a financial statement and offered to get the gas and build the pipeline within a few short months. I laid a certified check for $10,000 on the Mayor's desk and said, we'll forget this if we don't do just exactly what we said we'd do.

"Of course this was pretty hard on the competition because they didn't have $10,000. So we got the franchise and named our company the El Paso Gas Utilities Corporation."

As for the major pipeline financing, the new company obtained firm contracts with nearly all the gas users including Texas Cities Gas Company, which would convert its distribution system throughout the City to natural gas instead of the manufactured gas, a product derived from coal.

Only the American Smelting and Refining Company held out. It had been operating one of the world's largest custom smelters in El Paso since 1887, and they argued that gas could not be used to smelt copper ore. Their officers believed that only the white-hot flame from fuel oil would do the job, and that Kayser would never change their minds.

So Kayser went to New York and conferred with the general manager of AS&R. He also approached a brilliant engineer named Ernest de Coriolis, an experimenter with other industrial operations using natural gas and one who could design a gas burner capable of smelting copper ore.

Kayser propositioned AS&R's management offering to experiment with their furnace in El Paso. "If we can convince your engineers that natural gas is practical and efficient," said Kayser, "then I want to ask you to give us a contract.

"AS&R agreed, and in a short time we had that burner adjusted so that it was smelting copper better than it had ever done, and at a fraction of the cost of oil.

"True to their word, AS&R promptly signed the contract and converted to natural gas. Without AS&R, as conditions got tough, we would have had a lot of difficulty."

In September, 1928, Kayser rented a fourteenth floor office in the First National Bank Building. Previously, he had no desk in El Paso—just himself and what he carried inside his pockets and his head. Next he called the El Paso Business College and requested a stenographer. They sent over Miss Catherine Martch, a young girl who had just completed the course.

"She was the one for me," smiled Kayser. Catherine Martch was to be the Company's first employee, and she rose in rank to the office of Assistant Secretary-Treasurer, completing forty-two years of continuous service before she retired in 1971.

A few weeks later, the office moved to the seventh floor to more spacious quarters —three rooms. Kayser and Frost shared a double desk. A year later, in 1929, the stock market crashed. The great American dream was crumbling from its foundations. According to Kayser, "We knew the First National Bank was going broke and we didn't want any of the blame for them going under. We left a sizeable amount in the account, not wishing to contribute to a run on the bank. We hated to think of losing our deposit, but we couldn't stand the bad public position we would have been in, otherwise."

Ultimately, the First National, as well as a number of other banks in El Paso, failed, but customers received a small percentage of their deposits.

In spite of crumbling financial disasters, the just launched company stayed afloat and even moved to the recently completed brand new Bassett Tower. El Paso Natural rented half of the tenth floor and became the first tenants. Their vault and safes had to repose in the lobby for two months while the elevators awaited an okay by the state inspector.

Meanwhile, in September, 1928, crews surveyed the route of the planned pipeline —204 miles from Jal, New Mexico, to the City of El Paso. The pipeline company was incorporated in Delaware on November 22,

1928 as El Paso Natural Gas Company, and its directors were, besides Kayser and Frost, J. W. Colvin, A. C. Howard, W. J. K. Vanston, C. M. Weld and R. S. Jarvis. Kayser took over as President, Frost was made Vice President and General Manager and S. H. Benbow was Secretary-Treasurer.

In late December, the investment firm of White, Weld & Company financed the fledgling firm for $6 million and construction of the first pipeline began on January 5, 1929.

Shortly after construction started, it was obvious that someone was needed to ramrod the job from start to finish — a hard-driving, tough engineer who could get the line finished in the short six-months time promised by Kayser to Mayor Thomason and the City fathers of El Paso. Kayser found his man at Southwest Gas Utilities. His name was A. L. Forbes, Jr., an engineering graduate of Texas A&M University who became general superintendent and was destined to become legendary among El Paso Natural's employees.

Only Paul Kayser and very few others ever called him anything other than *Mr.* Forbes. Kayser called him "Todda." "I recognized that he was a man of great capacity," says Kayser, "although he was working for J. W. Colvin at Southwest Gas. I persuaded Colvin to let me put Todda in charge of the operation. I felt at ease about the thing when he took command. I knew that it would be done right."

Smith Brothers, Inc., a Dallas construction firm, had a contract for the sixteen-inch pipeline which was built from Jal, New Mexico, in a westerly direction toward Clint, Texas. It then turned northwest up the valley to El Paso. Brokaw, Dixon and McKee, an independent engineering firm, supervised and inspected the line.

The sixteen-inch pipe, made by A. O. Smith Corporation of Milwaukee, in thirty-foot lengths, was shipped by rail to El Paso and hauled to the job site in heavy World War I surplus trucks. All supplies, including water, had to be transported. Along part of the right of way there was insufficient natural growth to build fires for bending the pipe, making it conform to the terrain, and the pipeliners used old automobile tires as fuel for the hot-bending process.

The country between Jal and El Paso was rugged and lonely, with no telephones, no railroads or highways. Work crews passed only one house. At the Guadalupe Mountains, a hundred miles east of El Paso, the line crossed over the rim rock and plunged 500 feet down a forty-degree slope to the valley floor.

It was a rugged, exasperating new way of life for crews that had been brought in from East Texas. In the eastern part of the state a number of pipelines had already been laid; but these, for the most part, were put down through cotton fields and flatlands where the earth was comparatively soft.

West of the Pecos, pipeliners found only a hellish mixture of solid rock mountains, deep powdery sand, rattlesnakes and salt flats. And what little vegetation existed was covered with thorns that ripped many a pair of coveralls and left their calling cards

EL PASO in the late 1920s was the only major city in Texas not supplied with natural gas.

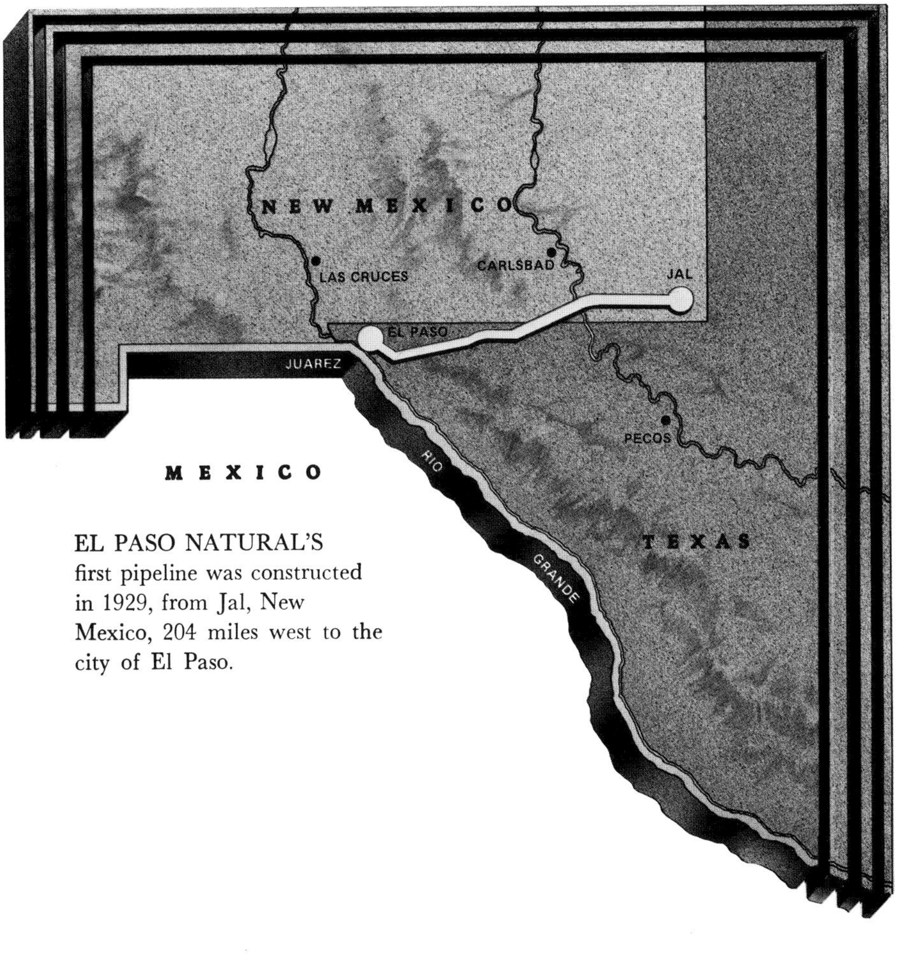

EL PASO NATURAL'S first pipeline was constructed in 1929, from Jal, New Mexico, 204 miles west to the city of El Paso.

A. C. Martch

HORSEPOWER, in the literal sense, was still used by
El Paso Natural in early-day construction of pipelines and
compressor stations.

underneath. Seldom, if ever, was there a
hotter, harder, dirtier set of circumstances.
The dirt roads of the late Twenties created
major obstacles to hauling men and equipment from supply points such as El Paso,
Van Horn, Sierra Blanca and Pecos. It was
a classic conflict between man and nature,
and it seemed at times that nature would
come out the winner, especially during the
spring sandstorms of 1929. Violent winds
tore the canvas-topped cook shacks and
tents loose at the foot of the Guadalupes
and blew them somewhere west of Fort
Worth. Dust and sand filtered into everything including machinery and bed rolls.

Even the flour and sugar turned brown.

This kind of pipelining was not yet in the
books. No previous pattern existed, and
numerous innovations marked the fitting of
more than 200 miles of sixteen-inch steel
pipe into one continuous piece from Lea
County, New Mexico to El Paso.

Most of the four-foot deep pipeline ditch
had to be ripped out by sheer back-break-
ing manpower, since heavy machines and
large tractors had not yet replaced teams of
horses and mules. In steep, rocky areas the
small tractors could not get a foothold in
the miles of broken boulders and slippery
shale. The steady footing of horses and

48

DURING THE DEPRESSION, the Company made some of its construction machinery take a back seat to hand labor and horse power. This furnished jobs for people along pipeline routes.

mules did the job. It was pick and shovel and jackhammer and dynamite most of the long, grueling way.

In spite of some of the orneriest, toughest terrain the Southwest had to offer, construction on the Jal to El Paso pipeline went on through the spring of 1929, right on schedule.

Along with the pipeline construction, a contract crew planted telephone poles in the hard earth and strung a wire line from Jal to El Paso to provide instant communication from one end to the other and from points in between. This wire line served for many years, not only as a hookup between the Lea County gas fields and city of El Paso, but it provided a vital link to civilization aiding stranded travelers, accident victims and people in need of medical attention. It was the only means of quick communication across the lonely desert stretching eastward out of El Paso towards Carlsbad, New Mexico.

By June, the Company's field system totaled six miles of pipe, and five wells were scheduled to supply gas for the pipeline. They had been tested and reportedly were producing a high grade of sweet gas which could be sent to El Paso direct from the wells without compression or treatment.

BY TODAY'S
standards El Paso's
pipeline methods were
primitive during the
formative years.

ORIGINAL SURVEY
gang on the Jal to
El Paso telephone line.

50

MAIN STREET in Jal, New Mexico in the early 1930s was usually ankle-deep in dust but became a morass of mud during summer rains.

IN 1929 hemlines were up, waistlines down. Well-dressed flappers were shopping for frocks like these advertised in the *El Paso Evening Post,* that prosperous summer before the stock market crash.

Sweet gas (with low sulfur content) was the only kind of natural gas being commercially used. With great quantities of sweet gas available in other areas, it was used by cities and industries elsewhere. Sour gas, which would require treatment to remove its sulfur content was considered unprofitable — so unprofitable in fact, that no one had ever designed or used any equipment to treat it in commercial quantities.

According to Paul Kayser, "one of the wells was tremendous, one of the biggest in the whole country, with fifty to sixty million cubic feet a day open flow. We started out with the sweetest, finest gas you could want; and I actually went out there with the engineer who took a sample of the gas for analysis to find out whether or not it contained any sulfur to amount to anything. I was with him when he turned the valve. I wanted to be sure. He took the sample and it showed sweet."

Then disaster struck. When the new Jal-El Paso line was being purged and loaded with fuel for the first time, the tell-tale odor of sour gas was detected. A quick check disclosed that three of the five wells had somehow begun producing sour gas. The Texas Production Company's Rhodes and No. 1 Cagle wells were sweet. And the No. 1 Cagle turned sour within the year. For the next four years, the Company had available sweet gas from only one well, since all the wells not then tied-in to the system were also found to be sour.

"The Rhodes well was always sweet,"

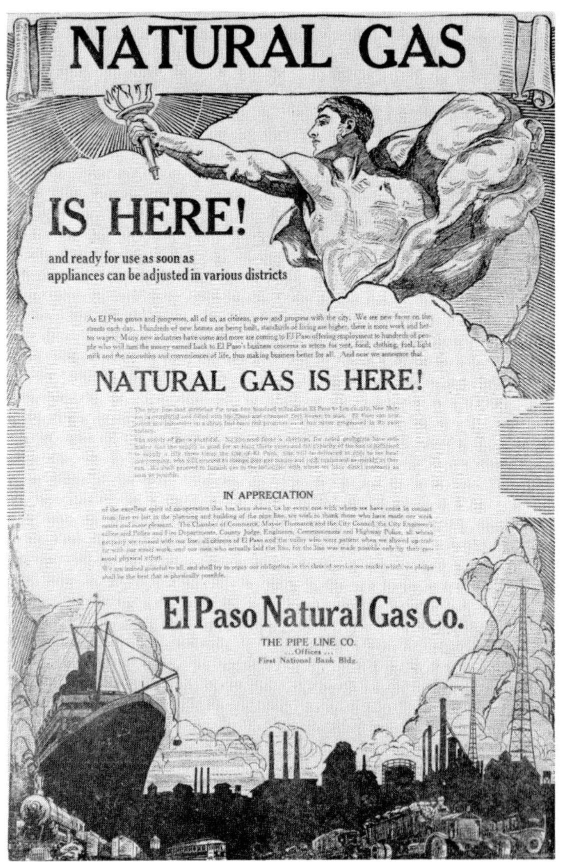

says Kayser, "but it couldn't deliver anything like the quantity of the other wells."

The new company was in deep trouble with its first setback. The product, for the most part, was not fit to burn. A. L. Forbes rushed to Los Angeles to enlist the aid of an outstanding engineering firm specializing in the design and building of refineries and industrial plants. Together they devised a method to treat sour gas, and experiments using a soda ash treatment started within days.

The first treating plant, Jal No. 1, purified up to about three million cubic feet of gas a day. But above this quantity, production dropped, and since the system had a total capacity of ten million cubic feet a

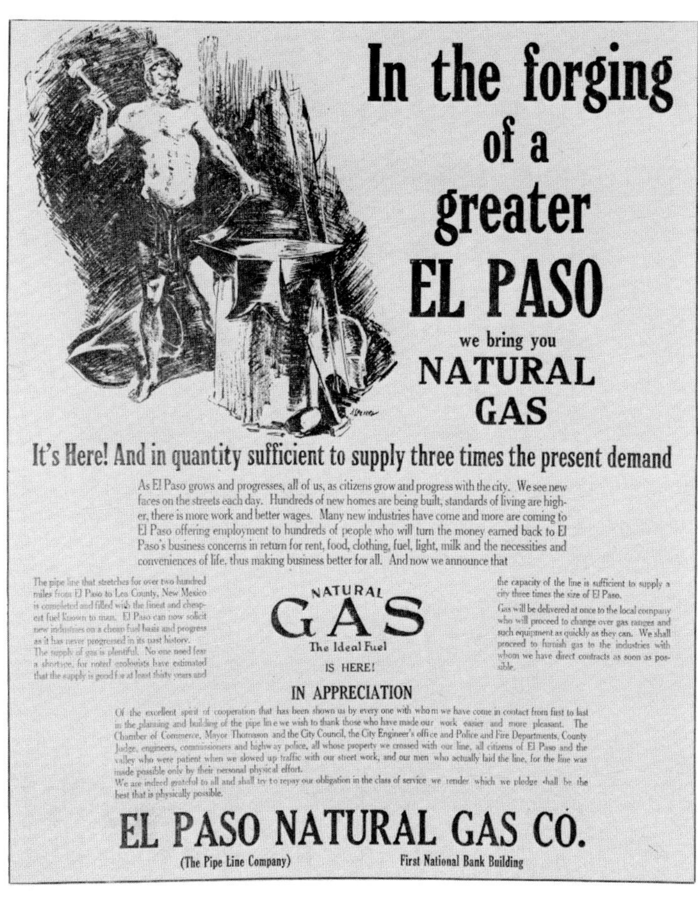

THE ARRIVAL of natural gas made headlines in El Paso. These advertisements appeared in El Paso newspapers on June 18, 1929. The occasion was celebrated that night at the Company's City Gate with Mayor R. E. Thomason and others participating.

day, the experiments continued. The Company literally worked in the unknown, designing and building its own equipment. The early treating vessels used soda ash as the basic chemical and brought impure gas into contact with it, a pretty primitive process. But it actually worked, and El Paso Natural climbed out of a very delicate, messy situation.

On the night of June 18, 1929, El Paso citizens devoted most of the evening to fanning themselves on their front porch. But several hundred people gathered at El Paso Natural's City Gate, six miles east of town to watch the official proceedings as natural gas made its first appearance in the City. The Municipal band played while Mayor R. E. Thomason fired a roman candle which ignited jetting gas into a giant flare. Flaming sixty feet into the hot night air, it signaled watchers for miles around that natural gas was flowing from over 200 miles away to serve local homes and industries.

First deliveries were made the next day, June 19. It flowed through one of the first long-distance, all welded steel pipelines ever built for natural gas. The sixteen-inch steel tubes had been joined, by welding, into one continuous piece of piping — clawing and curving its way 204 miles to El Paso from the tiny village of Jal, New Mexico.

A new name appeared on the payroll: C. L. Perkins.

THE JAL CONNECTION

Pipelining and bootleg booze

A YOUNG FELLOW named Chris P. Fox (who later became sheriff of El Paso County, a respected historian and a vice president of the State National Bank) delivered part of the pipe and supplies to the Jal-El Paso construction job. "I was in the oil field hauling and rigging business," says Fox, "and I had just finished doing a lot of work for Standard Oil Company, which came into El Paso as Pasotex, with their El Paso refinery and pipeline down to Wink and beyond.

"I was going down the street one day and got the buzz that things were beginning to pop with the proposed gas line from Jal to El Paso and a new company had just received a franchise from the City to bring natural gas to this area. So, I did a little rooting around and found they'd established offices in the Bassett Tower. The one who I first recall having met was Catherine Martch. Pretty soon they began to give me some work, mostly hauling pipe and supplies. And that's where a wonderful relationship started, wonderful as far as I'm concerned, and has continued through the years.

"One of the highlights of that early time, I usually recall, was the pusher-upper, the fella that got the job done — A. L. Forbes. Some people called him Todda Forbes. I didn't know what his official capacity was, but I know he was a ramrodder.

"One night I got a call about one or two o'clock in the morning. They told me to get down to the old metering station in Ascarate, which was really their operating headquarters. They said they'd had a break in the new pipeline. At that time they had it on wooden structures across the salt flats, elevated because they didn't want the pipe to be corroded by the salt. So anyhow, I broke out a bunch of rigs, everything I could get that would run — some ten or eleven of 'em — and we got down to Ascarate and loaded pipe. Todda Forbes was already there, running the whole show. He was a hard worker and he wanted to get a good day's work out of everybody. I don't say he was unreasonable. Those were hard working times, and if everybody didn't pull his share of the load, things

CHRIS FOX'S trucks unloaded heavy equipment for El Paso Natural in Jal, New Mexico.

bogged down. He didn't ask you to do anything he wasn't ready and willing to do himself.

"We loaded up and took off with those trucks of pipe, but we couldn't go out on the Carlsbad road; there was no real road then anyway, just a dirt trail and the Standard Oil pipeline road was washed out because of some heavy rains. We had to go around by way of Pecos to come up to this break on the east side, but we got the pipe on the job and a day or so later gas was coming into El Paso.

"Some years before the El Paso pipeline road was built, there used to be a dirt road from El Paso to Crow Flat. That was when they put together the El Paso-Artesia mail route. A Model T Ford went from El Paso, through Fort Bliss, then followed the old stagecoach route, up the canyon there by Helm's Ranch. It cut through Cerro Alto Canyon and turned sharp right towards Salt Flat, then to the left on the winding old wagon road over and through the Sacramentos, and then down into Artesia. It was rough going, I'm telling you.

"But those were good days and good things were done. And a good platform was put under our community and our country around here. I can tell you one

thing about El Paso Natural, even at that day and time. They didn't know it, but they were ecology conscious. Whenever we'd do any work for them somewhere, we were told to do a good job of cleaning up after ourselves. And no matter if it was in El Paso or out at Salt Flat, or where it was, we left a presentable appearance wherever we went, even when we were unloading around Sierra Blanca, with the gentle zephers blowing, we'd clean up and leave everything just like it was — maybe a little better.

"We hauled all kinds of oil field equipment to Wink, Odessa, Pyote, Monahans and the other towns down in the Permian Basin. We did a lot of hauling from the railroad at Pyote to Wink. Even though the road started out narrow through those gypsum flats, after you'd been over it with ten or twelve trucks, it was impossible because it was all chewed up with the gyp and full of big holes. Everybody just drove further and further out on the side of the road till it was a half a mile wide.

"I remember a restaurant in Pyote in those days that was really something. I guess it was easily a half-block long and maybe fifty feet wide. And a long counter with stools all the way down. All they served was steak and potatoes and coffee. One dollar. If you came in for breakfast, you got steak and potatoes and coffee, and it was the same noon and night, and one dollar. And I tell you the steak was huge, like a saddle blanket, but you ate quite a bit then because you consumed a lot of energy with the hard work and long hours.

EL PASO CITY GATE
as it looked in 1929. This was
the scene of a ceremony
performed by the mayor when
natural gas made its entry
into the city. Mayor R. E.
Thomason ignited a sixty-foot
gas flare on June 18 which
signaled watchers that gas was
flowing to El Paso from more
than 200 miles away.

"Of course all this was during prohibition days, and the bootlegging that
went on was really something else. Before we got hep to it, we must have hauled
an awful lot of whiskey down to the oil fields. These big boxes, you know, we
didn't pay any attention to what they were, but one day we began to smell a
mouse. We didn't smell any whiskey yet, just a mouse. I went to an old friend of
mine, one of the custom inspectors, and told him I thought we were hauling a lot
of things that were not oil field equipment. One evening he came by and said,
'Well, you've got some stuff here on your loading dock. Don't say anything to
anybody, you just let it go along and no doubt you will get a call from your driver
and he'll be all upset. But don't let that bother you, we'll take care of that.' Which
they did.

"What they found out was that the bootleggers were making heavy boxes
about four feet square and lining them with tin. Absolutely waterproof. In those
days, a case of Mexican whiskey bootlegged across the river at El Paso cost about
a dollar a quart bottle. In the oil field it brought about ten dollars. The box of
liquor would be delivered to a loading dock at some mercantile store or pipe
supply place in El Paso, billed to Pyote or Wink, and we would unwittingly load
it on our trucks. The rum runner in Odessa, Wink or Pyote would have an
arrangement with a particular El Paso store — with the owners never knowing
anything about it. He would just dummy bill the box on through to a friend in
the oil fields. It would be unloaded and somebody else would pick it up.

3

THE BOYS

FROM CLYDE

Halfway across the broad shoulders of Texas, the little town of Clyde drowses happily just off Interstate 20, surrounded by the oak-dotted mesas east of Abilene. A creek called Pecan Bayou meanders through the big lacy mesquite trees that line its banks. This is farm country, and Clyde has been a supply center for nearby farmers and ranchers since it first came to life in 1881. The town was originally established as a camp for the construction gang building the Texas and Pacific Railroad. They built a post office that same year, named for Robert Clyde, the camp boss. By 1885, the town had two churches, two stores and a population of thirty. In 1915, it supported two banks, five churches and the population had grown to 850.

Although it was endowed with a certain small town charm, respected for its neighborliness, Clyde, Texas was still a whistle stop, hardly the kind of place that appeared on travel folders or in *National Geographic*. However, in the late Twenties and Thirties it developed a claim to fame of its own; it spawned a group of special young men who joined El Paso Natural Gas Company in its early years and became highly-respected authorities throughout the entire gas industry as innovators and pioneers in a business about which few people knew. Three of them went on to become top executives of El Paso Natural: C. L. Perkins, H. F. Steen and R. W. Harris.

Perkins was the first to arrive on the scene. The son of a banker in Clyde, he spent several years at Hardin-Simmons University in Abilene and at John Tarleton College before going off on his own to work for Lone Star Gas Company in the nearby oil fields of Ranger and Breckenridge. He hired on as a gang pusher laying gathering lines, a job well suited to his tall, strapping, well-muscled physique.

Much of the credit for forming the nucleus of El Paso Natural's future employees must go to A. L. Forbes, who was then Vice President and General Superintendent of ·the Company. He had an uncanny knack of surrounding himself with rugged but bright people on whom he could count to get the new company on its feet. Forbes hired C. O. Byrne, who became the Company's first pipeline superintendent. Shortly afterward, Byrne, impressed with Perkins' ability to get along with the tough oil field hands, decided El Paso Natural Gas Company needed him.

Perkins smiles broadly as he remembers how he happened to come to work for El Paso Natural. "I'd been over at Abilene where they were having a coaching school, and I made arrangements to go back to school at a junior college; and if I made it, then I'd go to LSU if I was good enough. You got your room, board and your uniform and all that stuff. One night ol' Clarence Byrne called me—it was the first phone call put out of the Jal telephone exchange. He told me he'd give me $175 a month. And I said, 'goddamn Clarence, I'd sure like that, but the thing is I've got a chance to go back to school on a football scholarship.' He said, 'Well look, if you're goin' back to school to play football, get the hell out of here.'

"I reported to work on July 16, 1929.

"The Company's entire headquarters in those days," recalls Perkins, "consisted of three buildings about four miles south of Jal; one housed the office, lab and bunk room. Another housed the dining room and living quarters for the cook. There was also a combination stable and garage."

Jal was still not much of a town, although discovery of oil and gas in Lea County created a flurry of activity and brought in a handful of tiny service businesses, several cafes, bars and gas stations.

The coming of El Paso Natural Gas actually lit the flame that changed Jal from a village into a boom town of sorts.

Jal's beginnings go back to the mid-1880s when the Cowden Brothers Cattle Company drove a herd of rangy, rawboned Texas longhorns across the state line to stock their New Mexico ranch. Tradition has it they bought the cattle from John A. Lynch, an early-day Texas cattleman whose brand initials flanked the unruly herd. Their destination was Monument Draw, the ranch headquarters about six miles northeast of the present-day site of Jal.

It never really occurred to anybody that the Cowden brothers had just brought the name of a town across from Texas. The JAL brand survived the move—and change of ownership—simply because the Cowdens decided that rebranding the stock was too much work. They kept the brand and registered it under their name in New Mexico. The brand itself actually appeared

MR. AND MRS.
C. L. Perkins near the
Pecos River in 1931.

MILES FROM the
nearest town, El Paso's
pipeliners lived in
mobile tent camps
during early construction.

JΠL on the cattle, since leaving off the crossbar on the "A" prevented concentration of branding heat in one spot and protected cattle against infection. John A. Lynch, (if he was indeed the man for whom the town was named) in all probability, never saw the town because it did not exist until after the turn of the century.

The JAL Ranch flourished, and at one time it is estimated that more than 6,000 cattle were counted at one watering spot. The area covered by the ranch extended sixty miles north and south.

Homesteaders began to move into the Jal area about 1905, and it was not uncommon to see a JAL steer standing in the shade of a nester's shack. As the years passed, new homesteaders trickled in, and the ranch suffered a rapid decline. The open range was gone, and by 1910, the once sprawling JAL Ranch had dwindled to almost nothing.

A store and post office which had been opened earlier near the ranch headquarters were moved to the present site of Jal in 1916, and the town was born, officially, but unostentatiously.

The town site was originally a windmill-watering place established by the JAL Ranch. Ranch hands had dubbed the place "Muleshoe," but this name survived only a few years. When Charles W. Justis, a homesteader-turned-merchant moved his store to the site, it was desolate and remote, with no graded roads and no nearby towns. A few hackberry trees near the windmill provided the only shade for miles around. Justis' sons brought in the mail by horse-back from Monahans, some forty miles south, through sand, mesquite and catclaw. Supplies were hauled in from Midland and Pecos by wagon — ten days for the round trip to Midland and five to Pecos.

Soon after the store and post office were moved to Jal, the citizens established a one-room school house, and five pupils enrolled. By the early 1920s, the school had grown to accommodate thirty-five pupils, and Jal seemed to be on the map to stay. But homesteads which had been filed earlier in the area proved to be too dry for farming and too small for stock raising, so the settlers gradually abandoned their homes and moved.

Perhaps Jal would have dwindled and died completely except for the discovery of oil in Lea County in 1926. The little community, by coincidence, was sitting on top of some very prolific Permian Basin oil pools. The No. 1 Rhodes Well, six miles to the south, was brought in by the Texas Company, and Jal blossomed overnight into a town of 700 people.

Through the Thirties Jal continued to grow, although the Depression ended the hammer and nail boom. Oil and gas companies replaced cowpunchers and farmers, and Jal became the headquarters for geologists, engineers, roughnecks, drillers, and El Paso Natural Gas Company.

The next immigrant from Clyde to Jal was nineteen-year-old Hugh F. Steen. He had finished high school in May, 1929 and was working for a Fort Worth contractor, building highways throughout Texas. Re-

HUGH STEEN and
Gus Kendrick at Jal Plant
No. 1 in 1931. Steen was
warehouse clerk in Jal, New
Mexico. Below, Jal
construction crew raises a
tower for the soda ash process
for purifying gas. H. F. Steen
is standing at right.

calling those days, Steen smiles and says, "We were building a road between Vernon and Quanah, Texas, when 'I received a letter from Cy Perkins asking if I would like to come to Jal, New Mexico and work for a new gas company which had started operations earlier that year. Both Perk and I had worked for the contractor during vacations for several summers. Perk graduated before I did and had gone to college for a year or so before starting work for Lone Star Gas Company in the field around Breckenridge, Texas.

"Perk knew I was a nut about engines and machines of all types, and as it turned out, the gas in some areas was turning sour and it looked as though there would be some machinery around the yard. He advised that the pay would be 50 cents an hour, ten hours a day, seven days a week.

"I was getting 37 cents an hour running a form grader on the concrete road job during the week and the same amount for operating a clam shell on Saturdays and Sundays. My boss offered me a nickel an hour more if I would stay, but thank the good Lord I caught a bus for Jal, New Mexico. I finally got to Wink, Texas, the hub of a tremendous oil field area which was declining in production at that time causing the oil companies to explore the surrounding territory for new production. They did discover new oil as well as the natural gas area in New Mexico. I spent the night at the Rig Hotel as the Model A Ford bus only went over to Jal twice a week. My suitcase was lost on the trip from Vernon to Wink, but it finally caught up with me a week or so later.

"There was not much at Jal, but at 50 cents an hour working seven days a week, I thought it looked pretty good. Two buildings made up El Paso's Jal camp. One was the boarding house and the other was the office laboratory and bunk house combined. The utility system was a small Kohler light plant with output that would take care of six or seven 75-watt bulbs. The camp was some two miles from Jal. Wink was thirty miles away. Kermit was about halfway to Wink, but the only thing in Kermit was the Winkler County Courthouse and a store or so, along with very few houses. Eunice, New Mexico, about twenty miles north of Jal, was smaller than Jal, if possible, and Hobbs had not really started at that time. I well remember when oil and gas was found in the Hobbs area. It really boomed with numerous developments—Hobbs, new Hobbs, old Hobbs and surrounding shacks.

"From the start I enjoyed living in Jal. Things were rough and tough, but it was evident that we were into something new and different and that a little history might be in the making. Such turned out to be the case. The first winter we had some 200 breaks on the pipeline. We would go out to fix one and find two more before we got home a day or so later. Fortunately, at the time we were selling such a small amount of gas that parts of the system could be down a day or more and the line pack would still take care of the customers.

"The gas was sweet to start with, but after the wells were on the line for a while, many of them went sour and we became the specialists in devising and improving methods of purifying sour gas."

The third import from Clyde was R. W. Harris, a big lanky, raw-boned twenty-year-old kid who would play a major role in the future of El Paso Natural Gas Company. He came to work in June, 1930, less than a year after Steen.

Harris recalled that he had just finished a term at Texas Tech and gone home to Clyde for the summer. "Well," he said in his slow, central Texas drawl, "Perk called up his father, M. H. Perkins, who owned the bank there in Clyde and knew everyone in town. He wanted to know if there were any young boys around there who would like to come out to Jal that summer as El Paso Natural Gas had a bunch of wells in the field to tie-in.

"Mr. Perkins contacted me and said the Company was paying fifty cents an hour for laborers and wanted to know if I was interested. I agreed before he could finish the sentence, and the next thing I knew I was on my way to Jal.

"There were no railroads or paved roads to Jal in those days. I had never been too far away from home before. It was 275 miles, and when I got into Jal I figured it was the most far-away place I'd ever been. The chuckholes were deep and the roads were rough and the land all around was sandy with little scrub brush on it.

"The town itself consisted mainly of the Woolworth Hotel, a grocery store, a drug store, two pool halls, two houses of prostitution and any number of little shanties and shacks, and a lumberyard.

"People hauled everything out there by teams and wagons, even as late as 1930. But little by little the trucks started to come

R. W. HARRIS and James Bennett at Jal Plant No. 1. Harris was operator of the plant when this picture was taken in 1931.

through. In order to get through the sand, they would go in there and put caliche on the sand. But when the caliche broke through, you had these deep chuckholes and sometimes it would take you an hour to go ten or twelve miles. It was a common sight to see people stuck and broken down in hot cars in those days."

R. W. Harris literally started at the top and worked his way down — with a pick and shovel. "Everything was done by hand," said Harris. "There wasn't any modern machinery such as ditching machines, tractors and dozers. The work was done with teams of horses and crude types of equipment. The first job I had was helping put a distribution line into the village of Jal. Well, I had never dug with a pick or shoveled with a shovel. I just didn't know how. Everytime I would hit the caliche with a pick, the pick would bounce up and flop around, and my feet hurt and my back hurt, and I felt like giving up. I was beginning to wonder if I had made one of the biggest mistakes of my life by coming to the oil fields. Then I found out the trick from an old oil field worker. He said, 'Listen, Slim, let me show you how to use a pick and shovel.'

"He showed me just exactly how to take small bites and small chips with a pick and then neatly shovel it out. To this day, I am grateful to that man. If it had not been for him, I doubt seriously that I would have stayed with El Paso Natural for forty-two years.

"By 1931, they put us on half time and I was making the grand total of $75 a month.

I was in love with a beautiful girl from Clyde named Alice Bailey. Her father was the local doctor there. When Alice and I got back from Clyde after being married the first person we ran into was C. O. Byrne, and I will never forget what he said: 'Little lady, I don't know where you come from or what kind of a home you left, but you must really love this boy to come out here in these sand hills.' She didn't know whether to go back home or stay. We've laughed about that many times."

In 1930, when R. W. Harris was first learning how to use a pick and shovel in Lea County, New Mexico, the furthest thing from his mind was the general breakdown of the American economic system "back East somewhere." The effects of the 1929 stock market crash hadn't reached the sand hills of the Permian Basin, but nonetheless, the boom of the Twenties ended. It was only a matter of time until the gloom of the Great Depression was to spread like a great, gray blanket over the whole nation.

During the month of October the stock market began to waver; stock averages moved downward; yet President Hoover assured the country that business was on a sound and prosperous basis. But the market had a basic flaw; stocks were priced far above their real value, and speculators could buy them for a down payment of as little as ten percent from stock brokers who financed the rest. As stock prices slumped, brokers required the investors to put up additional funds. Since many stock speculators didn't have the ready cash to do this, they were forced to sell off some of their shares at low prices. This drove the market down even steeper. Economist Roger Babson warned that sooner or later a crash was coming and it would be terrific.

On Tuesday, October 29, 1929, Babson's prediction came pitifully true. The bottom dropped out of the stock market. Prices plummeted an average of forty points and speculators dumped their holdings in one stock to cover their investments in another. The value of stocks spiraled downward at a dizzying pace. The president of Union Cigar watched his stock drop from $113 to $4 on one day and promptly leaped to his death from the ledge of a New York building. Thousands of people lost everything; their life's savings and dreams of the good life were sucked into the whirlpool and disappeared down the drain.

But "Black Tuesday" in 1929 was not the real cause of the Great Depression, which had its beginnings in the industrialization and urbanization that started almost a hundred years before. By 1930, most of the country's wealth was in the hands of about thirteen percent of the people, and consumer goods were being produced at an unprecedented rate. Unfortunately, manufacturers of everything from automobiles to bathtubs were unable to sell off their mounting inventories to a population unable to afford this deepening sea of products.

Since the market was glutted with over-production, manufacturers began to close down plants and lay off workers. Some 75,000 men lost their jobs in 1931 when Ford closed his Detroit plant. The Depression sounded its depths in 1932 when one-third of the population of Pennsylvania was on relief. In Chicago, 700,000 people were unemployed. Thousands of Americans faced actual starvation and families all over the country managed to live on stale bread, thin soup and garbage. One company advertised jobs for 750 common laborers at twenty cents an hour, and 12,000 men showed up to offer their services. As millions of Americans tried desperately to withdraw their savings from banks, the whole financial system cracked wide open and by 1933, 5,000 banks had closed their doors.

As the Depression deepened in the heavily-industrialized East, it was only a matter of time until the farmers in the West were affected. Unemployed factory workers could not afford the beef and farm products being produced in vast quantities by American farmers. It was a stark fact of life that thousands of hungry tramps were roaming the countryside while food prices fell so low that Texas farmers were burning corn for fuel.

Unfortunately President Hoover couldn't stem the tide of mass poverty. The public heaped scorn on his head, although probably no one could have prevented the whole catastrophe. Americans were looking for a way out of their long dark tunnel, and the 1932 presidential election provided them with a chance to do something about it. Franklin D. Roosevelt campaigned vigorously for his "New Deal." "The country needs bold, persistent experimen-

MODEL KITCHENS showed the latest trends in gas ranges in 1930.

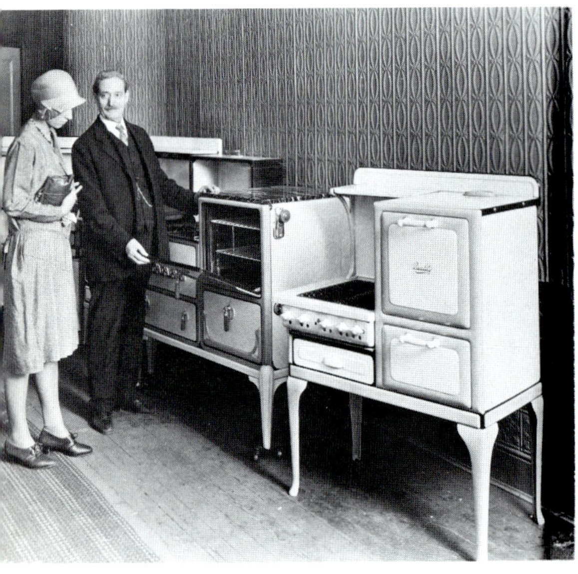

DISTRIBUTION WITHOUT WASTE

SAFEWAY Pay'n Takit STORES
Distribution without Waste   *All Over the World*
Prices Featured Jointly
PIGGLY WIGGLY

Continue to Save with Pay'n Takit

It is a pleasure to shop the Pay'n Takit way—Complete Food Stores located conveniently wherever you go in Arizona.

Prices Effective Wednesday, January 10 in All Phoenix and Salt River Valley Stores

LETTUCE — head **1c**
Extra-fancy Quality

Carrots, Spinach — Bunch **1c**
Turnips, Green Onions—your choice

YAMS 3 Lbs. **10c**
Fancy Arizona

APPLES 4 Lbs. **15c**
Rome Beauty—Northwest Grown

ORANGES Arizona Navals Large size, doz. **23c**

WHITE ONIONS 2 Lbs. **9c**

PEACHES
Our Choice — sliced or halves yellow cling. 2 No. 2½ Cans **21c**

MATCHES
Favorite — good strike — anywhere matches at a very low price. 2 Boxes **7c**

SNOWDRIFT
Pure Vegetable Shortening 3 lb. Pail **37c**

Syrup Max-imuM Cane and Maple Pts. **17¢** Qts. **29¢**

Ginger Ale Clicquot Club Pint **10¢**

White King Soap Granulated Large Pkg. **25¢**

SWEET CORN
Tendersweet No. 2 Can **7½c**

A-Y Cinnamon Rolls Pkg. of Nine **10¢**

A-Y Date Bread 16-oz. Loaf **10¢**

Baking Powder Calumet Double Acting. 1-lb. Can **25¢**

Flour Arizona Star and Rose 10-lb. Sack **39¢** 24-lb. Sack **83¢** 48-lb. Sack **$1.63**

Pork & Beans Van Camp's 2 No. 300 Cans **9c**

Baby Food Gerbers—vegetables, fruits, cereals. Can... **10¢**

Biscuit Flour STAR—ready mixed, add milk or water. Pkg. **27¢**

Cocoa Our Mothers'—a rich full bodied cocoa. 2-lb. Can **19¢**

TOMATO SAUCE Libby's or Valvita 2 8 oz. Cans **7c**

VINEGAR Highway—pure cider 12 oz. Bottle **3c**

Pickles Bread and Butter—Best Foods. 15-oz. Jar **13¢**

Quaker Oats Quick or Regular Large Pkg. **17¢**

TOILET TISSUE Your choice of Zee or Waldorf. 2 Rolls **7c**

Steaks
Shoulder Veal
Lb. **15c**

Veal Round Steak lb. **23¢**
Pork Back Bones 2 lbs. **9¢**
Loin Veal Steak lb. **19¢**
SAUERKRAUT Libbys' Barrel Kraut 2 lbs. **15¢**

Steaks
Loin, T-Bone, Rib
Lb. **17**

WE RESERVE THE RIGHT TO LIMIT QUANTITIES

DISTRIBUTION WITHOUT WASTE

SAFEWAY and Piggly Wiggly teamed up with newspaper advertisements in Phoenix touting Depression prices.

tation," said Roosevelt. "It is common sense to take a method and try it. If it fails, admit it frankly and try another. But above all, try something."

In the election, voters made Franklin D. Roosevelt the thirty-second President of the United States. In his inaugural address FDR reassured the American people with his now-famous phrase, "The only thing we have to fear is fear itself." Roosevelt immediately made sweeping changes in a system down on its knees and bleeding. Radios across the nation picked up the spark and blared the bouncy lyrics to "Happy Days Are Here Again."

But the problems remained not far under the surface. The darkest days of the Depression were still ahead and they were not to be shaken completely until 1939, the year German panzer divisions rolled into Poland to ignite World War II.

During the early Thirties, the Depression hit the oil fields like a West Texas tornado. With manufacturing plants shut down throughout the country and automobile plants almost at a standstill, the demand for oil and gasoline dwindled. The price of oil dropped to ten cents a barrel, paralyzing the petroleum industry. At ten cents a barrel, it wasn't worth the trouble to drill, and oil field workers left West Texas in droves. Author Sam Myres wrote that "families by the hundreds deserted their homes in McCamey, Wink, and Hobbs, taking with them in their rattling cars, only their essential belongings. Some of them found a measure of security with relatives who lived elsewhere in the nation.

Others took off for the oil fields of East Texas. An eerie silence fell over the oil towns of West Texas and southeastern New Mexico. The Depression closed in and held the region in its relentless grasp."

These were the conditions that prevailed when the budding El Paso Natural Gas Company was struggling to build a viable organization. The little pipeline from Jal to El Paso furnished the city with some six million cubic feet of gas a day, barely enough of a market to keep the Company afloat. Could it survive in the midst of the nation's worst economic disaster? Probably not—unless new markets could be found and pipelines built to supply them.

Paul Kayser decided to expand the Company—even in the face of the Depression. He would take gas westward from El Paso if he could find a market. This meant beefing up the Jal operation, tying in new wells, building a system of gathering lines to bring new gas into the primitive treating facilities in Jal. The word got around quickly, especially into the Depression-ruined countryside around Abilene, Clyde, and central Texas.

El Paso Natural Gas Company was hiring, and dozens of young men flocked to its Jal headquarters looking for work. Perkins, Steen and Harris hired many of them during those dark days, and these appreciative, hard-working men became the backbone of the Company. Their stories and impressions of their first days literally fighting for laboring jobs with El Paso Natural are filled with a certain bittersweet humor that seems unreal today.

One such employee was J.W. "Dub" Baulch, who now heads up the gas purchase group in the Permian Basin. "In 1931," Baulch recalls with a wide grin, "El Paso Natural was laying some pipeline out around Jal. That place wasn't over-populated then, and it wasn't too attractive to start with. But the old country around Clyde depended mostly on farming cotton, some ranching; no oil at all in that area, none whatsoever. And so there just wasn't anything for a young fellow to do. The Depression hit that country pretty hard. The people were just old country folks mostly, but they weren't afraid of work. I don't know of anybody that came from there that was afraid of a day's work.

"I'll never forget how I felt coming out to Jal. Perk said he could use some pipeliners. Fifty cents an hour. Well, I'd never made fifty cents a *day* up until that time. I'd also never been to New Mexico except once and that was up around the Clovis area. Gosh, when they started talking about Jal I could visualize a pretty green countryside, you know, like the mountains you always heard about in New Mexico.

"So I hitchhiked out to Monahans. You had to go to Monahans to get to Jal. When I got to Monahans I had to find out where the road to Jal was. I kept looking for a paved road, but all of U.S. 80 wasn't even paved in those days. So somebody showed me the depot and told me the road that goes north from there is the one you want to take to Jal.

"Well, I got out there and it was just two sand ruts. I stood there for a couple of hours and not a single car went by. So I walked back to the railroad station and got to talking to the agent. He told me there was a railroad that went all the way to Jal — once a day. They called it a Doodlebug and it was just an old converted railroad passenger car. They put a truck motor in the thing, no real engine or anything. The only passenger on it that day besides me was a little old lady. We left Monahans about ten in the morning and the engine began to get hot. The driver would stop and put water all over it. It would cool off and then we'd go a little further. We did this off and on all day until we got to Wink and then Kermit, and finally got into Jal about five that afternoon.

"Of course, I didn't know enough about the place to really ask any directions. I was just standing there and Steen came by in a pickup, so I hailed him down. He took me out to the No. 1 Plant and I spent the night out there in the bunkhouse. You had to have a little prestige and be on the payroll for a while to be able to stay permanently at the bunkhouse, so the next day I had to arrange for a place to stay in town. Steen and one or two others were staying in the bunkhouse. And that's all there was out there then.

"After we got there," chuckled Baulch, "I walked up a rock hill by the plant, and it was hot. In June, man, I'm telling you at 5:30 that afternoon it was *hot*. And I thought, boy, if there's any way I could get back to Clyde I would go.

"But I guess that the best thing that

JAL CAMP yard in 1930. Note tents in background.

GAS COMPANY cars navigated in rough country near the Guadalupe Mountains along the right of way between Jal and El Paso in 1931. At left is Jack Stricklin.

ever happened to me is when I stayed on. I've enjoyed every minute of it."

During those years the Company made every effort not to lay off any of its employees. Rather, they placed a number of them on a part-time basis so they would have some income, even though it wasn't much. The only plant was old Jal No. 1, using soda ash for treatment of the sour gas. The gas came straight out of the field lines, through the plant and on into the pipeline. And, as Dub Baulch says, "They kept the old No. 1 Plant going and gas going down the pipeline and everybody kept from starving to death."

Baulch wasn't getting rich on part-time work and decided that if he was ever going to get any kind of a formal education, he should get busy and go to college while things were slow around Jal. "So," says Baulch, "that's when I decided to go to college and I took off. When I got out of school I went right back to Jal. I had thought about coaching football, but I gave up the idea. There was something about the gas business that got in my blood.

"Sure enough, I got a job. R.W. Harris put me back to work and I'll never forget him. They were extending the ten-inch line from Jal to what is now our Jal No. 3 Plant. We were just north of town on a rocky, caliche hill; R.W. took me out to the job. Paul Wright had just gone to work for El Paso a short time before that and was pushing the gang out there. R.W. told Paul that I had just finished school and I was coming back to work for El Paso. Paul Wright—a little bitty fellow—said, 'Well,

R.W., I'll tell you, we'll just take old Dub and we'll give him a pick and shovel and see if he learned anything while he was in college.'

"That's where I went right back to; four years before when I'd started school, I'd laid that pick and shovel down, and then I came back and picked up right where I left off.

"But I've never regretted it. In all the years I've worked for El Paso, I think we've always experienced a sense of loyalty that you won't find in another company anywhere. And I think that it can be attributed to the management we had back in the beginning. That's Mr. Kayser, Mr. Cragin, Perk, Steen, Harris; people like that. An employee always got a lot of consideration no matter what his problem was. The boss was always ready to listen to you, and that meant a lot. It's a sense of security and it caused loyalty that you just don't see anywhere else. And I think it's one of the things that made El Paso Natural Gas the Company it is today."

Jack Stricklin, who later became a Vice President of El Paso Products Company and who is now retired in Colorado, was another of the young men who came to work for El Paso in its infancy. Stricklin's country-boy humor followed him everywhere during his years with El Paso; he never took himself too seriously. Or anybody else, for that matter.

Stricklin tilts his head back and laughs now about the things that happened in the Thirties. " 'Course they probably didn't seem too funny at the time," he said. "I remember my dad [L.M. Stricklin] started up a little ol' pipeline construction company at the worst time in the world. He had some money in the bank, but when that went broke, dad went broke right along with the bank.

"Dad knew Mr. Forbes and was lucky enough to go to work for El Paso in 1929. Well, he owed a lot of money and I was in school. Things were pretty rough so Forbes told him, 'Why don't you bring your boy out, put him to work and that would help.' So that's what happened.

"In 1930, I went to work for El Paso at what we called the old Van Horn Camp, located about fifty-five miles north of Van Horn, Texas. It sat right down at the bottom of Signal Peak. There's a valve set up there now. We were in tents and we had one little ol' corrugated iron shack where the foreman stayed. The Company had thirty employees at the time as I recall. The idea was I'd stay out of school a year and help dad 'til things got better. But they got a hell of a lot worse, and I never did get back to school. I was seventeen years old.

"Honestly, I don't know why people worked for the Gas Company back in those days, other than there wasn't another job anywhere. We had absolutely no fringe benefits, nothin'. There was no insurance, no savings plan, no pension, no vacation, no Christmas bonus. Listen, if you laid down your shovel you were out of work, because somebody else would pick it up, that's how bad it was. I really don't know why we stayed in the Company, because

later all of us had opportunities to go to work someplace else for more money when things got better. But maybe by then we were used to it. I just liked the people. Even Forbes. I was as scared of A.L. Forbes as I was of a seven-headed snake. But I loved him. That man, he cussed you out, and I'm telling you nobody in the world can even faze me after workin' for Forbes as many years as I did. He used a lot of adjectives. He was very descriptive.

"Mr. Kayser was just the opposite. I've known Mr. Kayser since I was thirteen years old. My dad was involved with him in this little gas company down in north-central Texas. I guess I just worshipped Mr. Kayser all my life."

All through the tough years of the Thirties the young men kept coming, lean and hungry and eager to find an honest day's work. And they weren't all farm boys. Some, like Frank Galle, had a natural talent for anything mechanical, but unfortunately his skills were generally unnoticed and unrewarded. "Back there in 1934," says soft-spoken Frank Galle, "I was running a little automotive garage in Sweetwater, Texas. Just barely making a living. My brother, Adel, had gone out to Jal earlier and worked for about six weeks on a pipeline for El Paso Natural Gas. But it was just temporary, and when they finished the line he came on back to Sweetwater.

"Later on as times got worse, Adel convinced me that we should go on back to Jal. He was sure we could get some kind of a job out there. So I just loaded up my mechanic tools on my car and we took off.

"When we got to Jal we set up a little tent and started 'baching.' We didn't even ask anyone if we could pitch a tent out there, we just found a likely spot and set it up. Lots of other people were doing the same thing. Then every morning we drove out to El Paso's office to check in and see if they had a job for us, and we did it for about four weeks with no luck. Most days there'd be about 200 people standing there in front of the Company's office trying to get a job.

"Jal was quite a little boom town then. There'd be a lot of people in town one day and the next day it seemed like they were all gone somewhere else looking for a job. So I went right out there the next morning and checked in with John Newberry. He told me they didn't have room for anybody else, and I headed on back to town feeling pretty low.

"I happened to go right by where they were working in a clay pit, and they had a little old tractor there digging up this clay and they were loading it in a dump truck with shovels. The clutch was slipping and they were having an awful time, so I stopped and talked to John Newberry. 'John,' I said, 'why don't you tighten up the clutch.' He said, 'What do you mean tighten up the clutch?'

"So I said, 'Well, I think there's a way to tighten it and fix the thing up so it will work right.' John asked me if I thought I could do it and I said, 'I'll sure try. I think I can.'

"So I got my tools out and in about ten minutes I figured it out. They cranked it up and the thing just took off. Ol' John was so delighted, he told me to go back to the office and tell them to put me on the payroll. I was tickled to death to get on at that time, even though they told both Adel and myself they'd only have about six weeks work for us and that'd probably be all. Sure enough, in about six weeks we went out to work one morning and they laid all of us off that had been hired for that special little job.

"We went back to our tent and were just sitting there wondering what we were going to do next. About this time, Paul Wright came riding by in his pickup and he said, 'Say, I got a little job out there at the plant and could use you boys for a bit. Would you like to go to work?'

"I told him, 'Man we sure would!' So we went out to the plant right then and went to work. And haven't missed a day since. I give Mr. Wright a lot of credit for my being with the Company today. He was really nice to me."

After three months of steady wages, Frank and Adel Galle felt extremely solvent, at least solid enough to move out of their tent. They pooled their savings, built a little two-room shack, and Frank went to Sweetwater to bring back his wife and little girl. "We built it on the Company's property," says Frank, "because about five other people had done the same thing and nobody seemed to mind. The shacks were right where the Jal No. 1 Plant is today. We even tied-on to the Gas Company's water line and they furnished us with gas and electricity."

The lean years of the Depression dragged

on endlessly throughout the whole decade of the Thirties. John Sparling, who now runs the Permian Division Warehouse, tells how it was as late as 1939. Born and reared in Gordon, Texas, some seventy miles west of Fort Worth, Sparling was walking the streets of Fort Worth looking for work. "I went to every company in town," he says, "and couldn't find a job anywhere. I'll never forget one interview I had with Superior Oil. I must have filled out a thousand applications and I kept going back to this guy and finally told him, you don't have a job for me and I'm just wasting your time. I'm not wasting *my* time, 'cause that's all I've got—time. So he said, 'Well, look at it two ways. There's always hope. I look at everybody that comes in here and I want to help 'em. You know, something might happen; there might be an opening that I don't know about right now. To look at it a little cruelly, somebody might die.'

"With that," says Sparling, "I gave up in Fort Worth, and as corny as it may sound, I took Horace Greeley's advice: 'Go West Young Man.' I went West. My brother in Gordon told me he knew this fella' that hauled butane out in the Midland area and he fixed it up so I could get a ride with him in his truck. I heard there was work around Hobbs and that's where I headed. On the way to Midland in this butane truck is where I met my first big sandstorm. I had never seen a sandstorm. We were on the outskirts of Midland on Highway 80. This ol' boy had a full load of butane, and that's pretty dangerous, you know. The static could blow you up.

I remember in some places where the sand was so thick you couldn't even see the radiator cap. And I thought, man, what a country! Cold, right out of the north, too. I don't know how much money I had. I didn't have much, just what my brother gave me, and I had my clothes. I didn't have a suitcase, I just had my clothes wrapped with a belt. I was traveling light, but I had to.

"Anyway, we got to Midland, and this ol' boy, I'll never forget his kindness, asked me, 'Where are you goin' to stay tonight?' I told him I don't know, but I'll stay someplace. And he said, 'You stay with me.' I said, 'Well, I hate to impose on you.' And he said, 'I don't care. You need help, and I might need help myself someday.' So we bunked together that night in a hotel somewhere in Midland. I couldn't find it now if I tried.

"He fed me breakfast that next morning and I thanked him. I wanted to get out to Seminole, then cut across to Hobbs. So he took me out to the highway and said, 'That's the road to Seminole.' And to Seminole I went. Somebody picked me up and gave me a ride. I stopped there and saw an oldtimer sitting on the porch of a hotel. I asked him about Hobbs and how to get there and what about the possibility of work. He said, 'Well, it's real slow here, and it's real slow in Hobbs.'

"I was pretty discouraged, especially when I got out to the edge of Seminole and discovered there was no paved road to Hobbs. Nothing but just sand dunes with a little old scrub brush and tumbleweeds. And it looked scary. I could see where the

"I spent eight hours trying to catch a ride,
that's how many people were
highwaying in those days,
hunting jobs or going somewhere."

road would go around the edge of those sand dunes, just curved, and I thought, a man could get scalped in country like that.

"I came back on into town, and I don't know how it happened but I had enough money to buy a bus ticket. So I bought a ticket to Hobbs. I didn't want to walk that trail up there, so we wound around the sand dunes in this old bus and I got to Hobbs. I spent the night there in a hotel. It was just a one-story building with a hallway that went from the front to the rear. There were probably fifteen rooms in all, and they charged a dollar for a bed. It was the kind of an oil field hotel that you could have seen back in Ranger or Breckenridge in the early boom days. Ranger was rough. They'd push you off the sidewalk and the mud would be ankle deep. Hobbs was strictly an oil field town, just like Ranger was when I was a kid and I lived there with my father.

"After that one night in the hotel I had seventy-five cents left. An ol' boy told me that the only outfit he knew that was doing any work in that area was El Paso Natural Gas Company. That's the first time I ever heard of El Paso Natural Gas Company. He showed me the right road and I got out to the edge of town and caught a ride. In those days it wasn't like it is now. Nobody was afraid to pick you up, nobody would pass you by. People would stand in lines for hundreds of yards waiting for a ride. And everybody was polite about it. You didn't muscle in and get a front spot, you went to the end of the line until it came your turn. I spent eight hours trying to catch a ride, that's how many people were highwaying in those days, hunting jobs or going somewhere.

"Finally I got a ride with a magazine salesman, and he took me right out to the Jal Plant. I put my roll of clothes on this old long loading dock that ran the entire length of the office building, and I, like everybody else, looked the area over to see where I was. You know, you had to get up your nerve to go in and ask for a job.

"I saw some fellas unloading a truck of soda ash and I heard somebody holler, 'Hey, Sparling, what the hell are you doing out here?' I looked and it was a kid I had gone to school with. We called him 'Empty Head Haney' in school. He bloodied my nose one time in a fight. Out here they called him 'Hair Oil' because he always wore his hair slicked back. So I said, 'Well Haney, I'm looking for a job, what do you think I'm doing out here?'

"It turned out I knew quite a few guys there, and thanks to them and C.C. Selley and Paul Wright they put me to work. There was no way in the world a man could have gotten a job in an out-of-the-way place like that without help.

"Some of the guys still kid me about getting to Jal with six bits to my name. And they say I've still got it."

Even at the tail-end of the Depression, the stories and odysseys of men trying to find a tiny place in the sun for themselves in the bleak job markets that still prevailed are enough to make a grown man cry. Indomitable men like R.H. Barnett, Rupert Kiker, L.E. McElrath, J.C. Williams and Paul Trout were just some of those that hung around the long Jal loading dock at Jal Plant No. 1 waiting to hear the magic word.

Barnett tells about walking out to the plant every day from Jal. Most people didn't have a car and they would walk through sand for miles to line up in front of the office. "They'd hang around out there 'til afternoon to make sure they couldn't go to work. A little man didn't have much of a chance in those days because they were looking for big boys that could really pick up a lot of weight and do a lot of shoveling. So if you were little you had to work harder than the big guys to stay on. If you kinda' looked up once or twice, well they had somebody else out there the next day. That's the way you had to work. I know we had one little ol' boy out there. He weighed about 150 and they called him 'Gopher Sam.' He'd fill about a half a shovel when R.W. Harris came around and he was just like a gopher—that dirt would fly! That's the way a lot of nicknames were developed back there."

J.C. Williams remembers that people literally lived in their cars right outside the gate to be sure they were available if a job came up. "In the cold weather they cooked on a campfire and slept in their cars, bedrolls, whatever they had. They stayed there day and night thinking maybe the Company would need another man or somebody'd quit or get run off. They didn't even ask what the going rate of pay was."

L. E. McElrath recalls the long loading dock where he spent the first day with the Company sitting on a suitcase. "I can remember the happiest words that I ever heard about that time when Steen told me to come out in the morning and get on the truck. Boy, I was there the next morning, still sitting on the suitcase before work time. R. W. Harris was down on the pipeline with a gang and that's where I was supposed to be headed. I was sittin' on the dock and man, a lot of people began to show up and took off in trucks. I kept sittin' and the trucks kept loadin' and the first thing I knew there was nobody around, and there I was still sittin' on that damn suitcase. I thought the world had quit turnin' when the truck and everybody left and I wasn't on it. I can still remember the feeling. But finally, Steen came by and said there was another job and I could get work for a day or two. And I'm still here."

Williams remembers his first day was on a Sunday. "They were workin' and said they needed a hand. But I had to go to Kermit first [Kermit was known as the chicken fried steak capital of the world] and get a physical and talk to the clerk. He worked on Sunday, too. So I took a physical, came back about three in the afternoon and told the clerk I'd see him in the morning. He said, 'No, go up there and see a guy named Baldy Henderson, 'cause he needs help on pouring engine blocks.' So I did and worked nine hours straight. No lunch, no gloves, no cap or anything else, but at least I was smart enough to come back to the office with that physical on Sunday."

Meanwhile, the little community of Clyde continued to be a springboard of opportunity for out-of-work men who were anxious to take on the rugged jobs and spare lifestyle provided by El Paso Natural in Lea County, New Mexico. Through the years, more and more names of boys from Clyde were added to the employee roster and most of them went on to become well-known and important figures in the Company's development and operations: T. J. Crutchfield, H. A. Jones, H. C. Doan, J. T. Baulch, Hayden Pyeatt, Roger Smith, Sam Damewood, Bobby Adams and numerous others.

But not all of them were enchanted with the Gas Company at first. H. C. Doan, who is now superintendent of El Paso's far-flung Southern Division says, "I went to work the first time in 1935 as summer help in Jal for R. W. Harris. He put me out there diggin' ditches in that sand and I didn't like it, so I quit and came back to Clyde. Then I decided to get married and I wrote Harris a letter tryin' to get back on, but I never got an answer from him. I guess he was still not too happy with me 'cause I had just walked off the job in those sand hills.

"Well, I went on out to Jal anyway and an ol' boy named Cotton let me sleep in the bunkhouse with him. It was a dollar a day and he paid my dollar and also bought my meals hoping Harris would give me a job. I stayed there about four weeks, goin' down to the loadin' dock every mornin' waitin' to go to work.

"One mornin' I went out there and Harris looked around and said, 'Is there anybody here who can climb telephone poles?' I held up my hand and said I could climb 'em. Lord, I'd never been up a pole in my life and I think R. W. knew that, but he sent me to V. V. Vincent who was combination telephone line repairman and line rider. We had to build a line from the town

of Jal to the No. 1 Plant. Anyway, when we got out of the pickup, Vincent threw the climber's hooks out on the ground and said, 'Put 'em on and let's go.' 'Well,' I said, 'you're gonna' have to show me how to put those hooks on 'cause I don't know how.' He was good enough to show me and I finally did learn how to climb poles."

T. J. Crutchfield, who later became superintendent of the Permian Division, also went to work in Jal in 1936 as a laborer. Says Crutchfield, "R. W. Harris was pushin' the gang and we went to work on the side of one of those caliche hills puttin' in a six-inch line to the old Shoales Plant. This was one of the first plants they had in the Gas Company, and they tore it down shortly after I came to work. Perkins was the superintendent and R. W. was the gang pusher. When I went to work for the Company I had been workin' for ten cents an hour, ten to twelve hours a day. That was the goin' wage around Clyde and I felt like I was doin' real good until I joined El Paso. They paid fifty cents an hour and boy, that was top pay!

"I worked on construction for a short time, and one mornin' Harris said he needed two men to go down and help Estin Scearce overhaul the engines at the No. 1 Compressor Station at Malaga. And then as an afterthought he said, 'Oh, you might take your clothes 'cause you'll be down there a week or two.' Well, it turned out I didn't get back to Jal until after the War in 1945.

"I never did have a car there at Jal, but once in awhile some of us would go over the dirt road to Wink which was still a boomin' oil field town and there was some night life. It was a tent town; tents everywhere and most of the people lived in either these tents or shacks. About our only entertainment in Jal was to go up to town and watch the fights. You'd go and sit around town and pretty soon they'd be havin' a fight in one of the bars. Then they'd move out in the middle of the street. But that was about the high point of excitement in Jal back in '36."

Living conditions for El Paso's early employees in Jal could hardly be called lavish. They were, in fact, pretty crude— even by Depression standards. But it didn't occur to anybody to complain. They were happy to have a roof over their heads and walls to keep out the stinging sand storms and they could thank one man — more than anyone else — for his generosity. His name: James Bennett.

Some years before El Paso Natural built its Jal headquarters, Bennett homesteaded 160 acres, most of it immediately south of what was to become the old No. 1 Plant. Unfortunately, although Bennett's 160 acres sat atop the Permian Basin's oil and gas wells, he didn't own the mineral rights and consequently never made a dime directly from the petroleum. Nonetheless, James Bennett was an enterprising individual, enough so that he made some bunkhouses out of his chicken pens and rented rooms to the new influx of El Paso employees.

The Company bought part of his property on which to construct the Jal Plant No. 1. R. H. Barnett remembers that "part

of the plant was built on Bennett's property, so Mr. Forbes built him a new house down there and hired him to work for the Company. Everybody called him 'Preacher' Bennett because he used to try to preach a little bit. People came in and just threw up tents and squatted out there on James' property. But finally, he began to sell off lots and Company employees bought them and created a little ol' shanty town with tents and tin houses. It soon became known as Bennettville, with a post office and everything. The post office was known as Bennett, New Mexico.

"Steen's mother-in-law, Mrs. Marshall, ran the post office and a little grocery store and we'd go down there and eat and get sandwiches and stuff. She was a real nice person, always wanting to help people in any way she could."

John Sparling remembers living in one of Preacher Bennett's rent rooms. "Nearly

FIRST FAMILIES to occupy cottages at Station No. 4 near Deming, New Mexico pose for a photographer in 1931. Note pistols casually displayed under belts of two of these gentlemen.

all the employees lived in Shanty Town and it later became just as good a town as Jal. We had a family-style cafe there and you could get three meals a day for a dollar. The cafe was owned by a man named Sam Larkin. And could Mrs. Larkin cook! And Mrs. Kessee, her sister. They were the two best cooks in the country. And they're still there in Jal, fine women."

By 1938, there were eighty-four houses in Bennett. Today, part of the little settlement still exists—just south of the fence at Jal Plant No. 1—with many of the same shanties built back in the Thirties partly hidden by the large Chinese elms that have grown up around them.

James Bennett retired in 1968 and still lives in the same neighborhood, not particularly concerned about what might have happened if he had owned the mineral rights to his original homestead. He was always one of those rare individuals more interested in helping people than in amassing a fortune. "Besides," says Barnett, "even if he'd had rights to the oil and gas, he'd have brought in chickens or something —and they would have all died.

"He decided to go into the bullfrog business one time. He drilled a water well and put up a windmill and ordered these frogs from Fort Worth or some place back East. He had a trench dug down to his lake, but the sand soaked up the water before it ever got to the lake, so he never got any water in it. He never had enough water to fill a dish to get into that lake. That's just the kind of guy he was."

Although the times were tough, the boys from communities like Clyde were still hanging in there. Great things were to transpire in the growth of the Company, but neither they nor the management had any way of knowing what the outcome of the decisions to 'stick it out' would be.

4

MOVING WEST

As THE ECONOMIC DISASTER of the early Depression days struck the Permian Basin, the death knell sounded for a number of small oil and gas companies. El Paso Natural Gas could have easily suffered the same fate. Although the little company made black figures from the very first — net earnings of $282,500 in 1930 — it became apparent to Paul Kayser and H. G. Frost that troubled waters were rising fast and could engulf them any day.

True, they reasoned, the Company might be able to keep from drowning, but if it were to be anything more than a speck on the map—an undersized operation making puny profit, it was absolutely necessary to expand. It was either "go" or "no go." Put new chips on the table or get out of the game. The closest markets were the copper smelters and power plants at Douglas and Bisbee, Arizona and Cananea, Mexico. Under normal conditions, these industrial loads for natural gas would support an extension of the original El Paso lines into southern Arizona and northern Mexico if sales contracts could be made with the industrial plants.

Kayser and Frost decided to put new chips on the table.

First they tried for needed new contracts. Kayser went to Douglas, Arizona and talked to P. G. Beckett, manager of the Phelps Dodge Mining Properties. "He listened to me carefully," says Kayser, "but he didn't want to be the first to buy natural gas in Arizona. He told me to get some other contracts to examine. And just as I was leaving he said, 'Whether you get them or not, come back and see me.'

"So," says Kayser, "I went to Saginaw, Michigan. I went to New York. I went to Los Angeles and San Francisco, and I got nothing. When I got back to Los Angeles I was sick and tired of the whole damn business. I called Frost and told him there was just nothing doing and that I was coming on in. We would go to New York and trade or sell the stock in the Company to anyone who wanted to buy it and be done with it.

"He was terribly disappointed, of course. I sat down and sadly began to review the situation and then remembered what Beckett had said to me in Douglas. It stuck in my mind—'You come back and see me anyway.'

"I jumped up and called Frost and told him I wasn't coming in yet; I was going to call on Beckett and talk to him again. I decided to give the horn one more blow."

This added effort became one of the most dramatic turning points in the story of El Paso Natural. Beckett arranged for Kayser to meet with the Phelps Dodge top management in New York. After that meeting, a jubilant Kayser walked out with the signed contract to supply fuel to Phelps Dodge.

That opened up other possibilities.

The other copper companies in Arizona, around Douglas and Bisbee and Cananea, Mexico, could see that Phelps Dodge would save several hundred thousand dollars a year in fuel costs. "They didn't want Phelps Dodge to have that kind of an advantage," says Kayser, "so we signed up the rest of them."

Because of the problems of getting additional financing for a westward extension of the pipeline during the Depression, Kayser and Frost incorporated a new company, Western Gas Company, to supply gas to southern Arizona. Western's contracts covered most of the fuel requirements for fifteen years or more for the plants located at Douglas and Bisbee and Cananea, Mexico—including the Phelps Dodge Corporation, Calumet and Arizona Mining Company, Cananea Consolidated Copper Company and Arizona Edison Company. Western would also later serve the towns of Las Cruces and Deming, New Mexico as well as Douglas and Bisbee, Arizona, and Cananea, Mexico.

The new pipeline would cost something over six million dollars, but Kayser could not sell the bonds, so the Company persuaded the contractors and material suppliers to take short-term notes in payment for their materials and services. Apparently the only thing in abundant supply was the product itself; there was plenty of natural gas in Jal, New Mexico.

Enormous financial obstacles blocked the way. It wasn't easy to convince Depression-ridden financiers that the 290-mile pipeline to southern Arizona was a sound business venture. This six-million dollar extension would increase the total length of the line from Jal to about 500 miles. Never before had a natural gas pipeline of such length been constructed. To be operated economically, it would have to be designed for pressures far higher than those ever before used to transmit gas for any appreciable distance.

Compressor stations would have to be built; two between Jal and El Paso, on the original sixteen-inch line, and two on the new twelve-inch line to Arizona. Plans were made to build Station No. 1 some seventy-five miles west of Jal and Station No. 2, 135 miles farther. Station No. 3 was to be

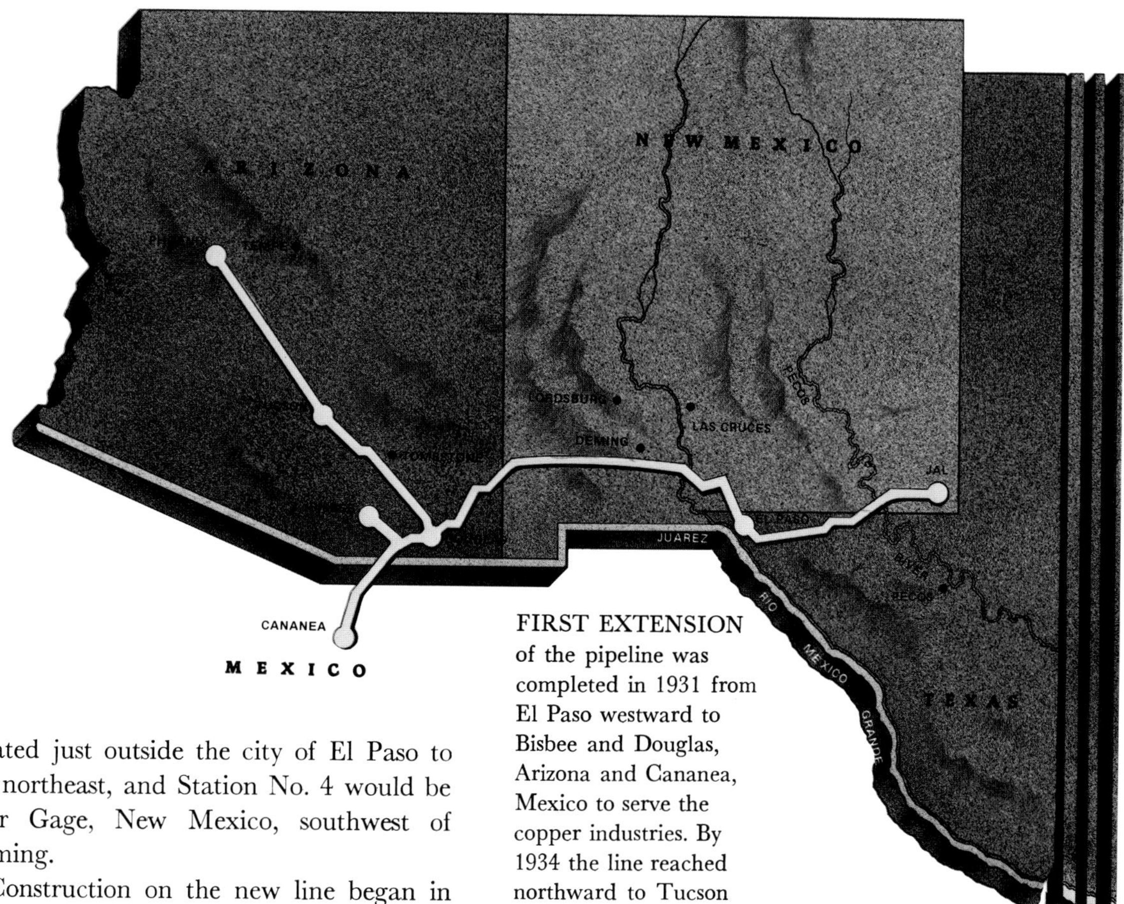

FIRST EXTENSION of the pipeline was completed in 1931 from El Paso westward to Bisbee and Douglas, Arizona and Cananea, Mexico to serve the copper industries. By 1934 the line reached northward to Tucson and Phoenix.

located just outside the city of El Paso to the northeast, and Station No. 4 would be near Gage, New Mexico, southwest of Deming.

Construction on the new line began in February, 1931, and was completed in June, two months ahead of schedule. The four compressor stations were all pumping gas westward by September. One of the new employees who worked on the project was a construction clerk named Virgil Rittmann, who would later become Secretary and Treasurer of the Company.

The Arizona pipeline represented a great advance in the natural gas industry, since it pioneered the long-distance transmission of gas and was, in effect, the granddaddy of the large-diameter lines that came later. With the turning on of service to Douglas and Bisbee and Cananea (through a subsidiary, Compania Occidental de Gas, S.A.) the Company roughly doubled its size. By adding southeastern Arizona, southern New Mexico and the northern part of Sonora, Mexico, El Paso transformed itself from an organization local in character to one of regional importance.

Morale of the Company's sixty-seven employees zoomed to new heights. Names that were to become familiar in later years appeared on the payroll. E. W. Woody, Roger Van Bramer, and others. John Schaffer led the survey party that plotted the new line. He hired a young engineer named John Eichelmann to go into Mexico to finish the survey on that portion of the job. Eichelmann was later to become chief engineer and Senior Vice President.

Estin Scearce, who retired in 1966 after thirty-five years as one of the prime builders of the Gas Company, said that "From the beginning, the new compressor stations were under the supervision of C. O. Byrne. Later a compressor department was set up with Walt Miller as its superintendent.

"C. O. Byrne hired John Crutcher, A. F. Pierce, Wilson Hodge and others from Lone Star Gas Company in Fox, Oklahoma. Walt Miller came from Humble Oil and Refining Company, and I was working at the Texaco Refinery in El Paso when I started with the Gas Company."

Enthusiasm and a feeling of buoyancy filled the air, more than enough for everybody. But no sooner had the Arizona pipeline been completed than the copper companies found themselves in deep trouble. The Depression had driven the price of copper down to five cents a pound, about half enough to make it worth mining. The copper companies might have to close down, a situation that would be disastrous for the new Western Gas Company, El Paso Natural's extension to the west.

Kayser reasoned that the way out of the situation was for the copper companies to get a better price for their product. "The only way that such an increase could be obtained," says Kayser, "was to get Congress to put a tariff on copper imports. Foreign copper, principally from the mines in Africa, was selling for five cents a pound, thereby setting the U.S. price."

The oil industry was also in trouble and was trying to get Congress to put an import tax on foreign oil. Kayser decided to join with the copper companies, the oil industries and a few others in need of protection from foreign goods in an aggressive campaign to get Congress to impose an import tax. The economy desperately needed all the help it could get to fight the Depression. Congress, fortunately, was sympathetic to the cause, and passed the import bill in 1932, placing an import tax of four cents a pound on copper, forty cents a barrel on oil and similar taxes on other products. The price of copper immediately rose to nine cents a pound and all talk of closing the mines faded.

Although the copper mines and smelters remained open, the Depression was so severe in 1932 that the country's demand for copper weakened to a trickle of what it had been. The copper companies cut their production to twenty percent of capacity, and as a consequence, the amount of natural gas used at Douglas-Bisbee and Cananea was not enough to support the new western line which was operating in the red. There wasn't enough demand for gas to keep the compressors working twenty-four hours a day. At night they were often shut down and the station crews handled routine maintenance work along the pipeline.

By 1933, the Depression hit a new low and there was no need for many of the employees then being carried on the payroll. The Company could easily have cut back on its personnel, but it still had faith in the future and held on to its people, waiting for the time when their skills and experiences would be needed. Employees worked on a half-time basis, a device that kept many of them from joining the ranks

IT ALL STARTED HERE

THIS LITTLE CLUSTER of buildings fighting for existence in a sea of fine, powdery sand once represented the heart of El Paso Natural Gas Company's operations. This was Jal Plant No. 1 as it looked in the early 1930s, within a year after the founding of the Company. The first two frame buildings at the left were occupied by employees and their families. The third building was used as a bunk house for bachelor employees. The Company's office for its activities in the Permian Basin was located in the fourth building. The house at the extreme right was occupied by Mr. and Mrs. C. L. Perkins, and the Paul Wrights lived next door. In the far background can be seen the soda ash treating vessels which were used to process natural gas before it entered the 16-inch pipeline to El Paso.

WORLD WAR I vintage
truck unloads 12-inch pipe to
be used on new line to Arizona.

DITCHING MACHINE
replaced the pick and shovel
on part of the westward
extension.

NEW COOPER-BESSEMER engines were installed in 1931 at Compressor Station No. 3 near El Paso to help move gas westward.

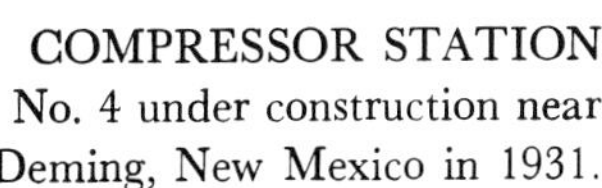

COMPRESSOR STATION No. 4 under construction near Deming, New Mexico in 1931.

of thousands of unemployed who sat glumly atop railroad boxcars roaming the country in search of other jobs.

Family men who lived without charge in houses at the compressor stations, considered themselves lucky to have roofs over their heads. There was a lot of hunting and fishing that year; the object was food on the table instead of sport.

For want of surveying work, John Eichelmann became a night dispatcher at El Paso City Gate in Ascarate. As Paul Kayser commented, "To put it mildly, this was an extremely difficult episode in the life of the Company."

Faced more and more with the problem of unused pipeline capacity, the Company decided on another expansion move that might help fill the partially filled pipeline— this time to Tucson and Phoenix, Arizona. A survey of the market showed that even in Depression times, the utility companies and small industries would support a small ten-inch pipeline from a point near Douglas to Tucson and Phoenix, and yield a clear profit of perhaps $300,000 a year. The major obstacle was that of financing another pipeline.

Under the new Roosevelt administration, the government had established a lending agency, the R.F.C. (Reconstruction Finance Corporation) to help finance companies that were in danger of bankruptcy as well as for profitable extension of their businesses. El Paso Natural applied to the R.F.C. for a loan of $2,200,000. Along with the application, Kayser called attention to the R.F.C. that the Tucson-Phoenix

area was costing the government more than a hundred thousand dollars a month in relief payments to people out of work. Kayser agreed to give these people an opportunity to work on the proposed pipeline at higher wages than ever paid before for this type of work in the state of Arizona.

The loan was granted, and construction by Bechtel and Kaiser on the new ten-inch line began in the late summer of 1933. Stone and Webster did the engineering and inspection. That year also marked the appearance of a new face in the management of El Paso Natural: Charles C. Cragin. Although he did not come up through the ranks as did many of the original pipeliners, Cragin brought to the Company wide experience as an executive and an engineer in the utility field. He was the former manager of the Salt River Valley Water Users Association in Phoenix and had a thorough knowledge of conditions in Arizona where El Paso Natural was rapidly expanding. Cragin came in as Vice President and assistant general manager. He played a major role in the Company's healthy growth until his retirement in 1953.

In the summertime, Cragin could be easily spotted because he invariably wore a white Palm Beach suit and a straw boater— even into the early 1950s when he retired. A true story Cragin told on himself concerned the time when he was just out of college, New York University. He had spent most of his life in New York, and it showed. "I was hired to go to work on a railroad construction gang. I got off at this station at the end of construction for the railroad and was standing there on the platform with all my belongings in a paper bag. An old resident farmer drove up and hitched his team, so I told him who I was and that I wanted to be taken to the camp. The fellow took one look at me and said, 'Christ couldn't come, so I guess He sent you!' "

On December 11, 1933 when natural gas made its entry into Tucson, the *Tucson Daily Star* proclaimed that "A huge private enterprise became, in effect, a relief agency for the many communities along its route." To provide more jobs, machines took a back seat to make room for hand labor on the Tucson pipeline. It was manpower which opened the trench for the laying of this 217-mile extension. Pick and shovel labor for the line was obtained from each county as the construction progressed, and these pipeliners were hauled back and forth to their own communities each day. In this way, the men on the job got the benefit of the pay, and communities in which they lived got the benefit of the construction payroll. Phoenix received natural gas on January 11, 1934.

This venture was destined to be the turning point for the Company. The new line was an immediate success and earnings showed a great increase in the next three years. By 1936, the violence of the Depression had subsided, and El Paso Natural refinanced all of its debt and paid the R.F.C. loan in cash. "After that, we had a *Company*," says Kayser. "We paid off everybody and put everything under one mortgage. We were as solid as a brick wall,

C. C. Cragin

LOWERING IN operation was accomplished with a chain hoist during the Company's first years.

JAL PLANT NO. 1, above, as it looked in the early 1930s. Soda ash process was used to treat sour gas in the Permian Basin. Below, aerating towers were landmarks for many years at the Jal Plant.

and from that time forward we were always able to obtain the money for any project we wanted to build." In 1935 H. G. Frost resigned his post and C. C. Cragin became Vice President and general manager. Frost remained on the Board of Directors through 1941. J. E. Franey became Secretary and Treasurer. On November 2, 1936, the New York Stock Exchange listed El Paso Natural's common stock for the first time.

From the time the first Jal to El Paso pipeline was built, construction of pipelines and plants had been contracted out to independent firms, but the unusual terrain and much of the Company's construction work soon indicated that a permanent force of workmen familiar with local problems would be a profitable addition to the employee roster. In 1935, the Company formed its own construction department with C. C. Selley heading up plant construction, and J. F. Schaffer supervising pipeline construction.

El Paso's tiny engineering group designed a new treating plant at the Jal No. 1 Plant, using equipment, designed and developed within the Company, which was unique in the industry. The time had come to put together an experienced force to expand and improve equipment such as this. It also seemed right that the new construction department could do the work of building pipelines and plants more efficiently, and so it did just that.

The pipeline construction department undertook its first project; forty-five miles of pipeline running from Casa Grande, Arizona to Superior for gas service to the Magma Copper Company. Summer temperatures of 118 degrees are not uncommon around Casa Grande, and the town is still referred to by many as "the fan belt capital of the world."

By the end of 1935, El Paso Natural had a total of 132 operating employees and a network of small new lines began to spread into the towns of Coolidge, Florence and Superior, Arizona. Officers of the Company in 1935 were Paul Kayser, President; C. C. Cragin, Vice President and general manager; A. L. Forbes, Vice President; J. E. Franey, Secretary and Treasurer; Frank A. Liddell, General Counsel; A. C. Martch, Assistant Secretary and Assistant Treasurer; T. W. Gentiles, Assistant Secretary; and S. H. Benbow, Assistant Secretary and Assistant Treasurer.

As the load on the line to Phoenix and Tucson increased, the gas originally transmitted without compression in this area, needed boosting. Compressor Station No. 5 was constructed near Douglas, followed a few months later by Station No. 6 near Benson, Arizona. New pipelines to Silver City and Hurley, New Mexico, and to Ajo, Arizona were completed in 1936. The right of way to Ajo passed through eighty-three miles of the Southwest's most rugged country, much of it through the Papago Indian Reservation. It was the Company's first experience with Indians, and the Papagos were understandably dubious about the whole project. To top it off, the weather was miserable—and John Schaffer remembers that construction workers were quitting in droves — almost as fast as they could be hired. The job was finally finished by roustabouts from a carnival which had wandered into the country. In 1937, pipeline crews built lines to Hayden, Arizona and Lordsburg, New Mexico.

While pipeline construction crews busily laid lines to new markets in Arizona and New Mexico, the Jal field headquarters tied-in wells and added field lines to supply an increasing demand out on the west end.

A. L. Forbes still ran the operation. R. H. Barnett recalls, "We'd hear that Mr. Forbes was on his way to Jal from El Paso and ol' R. W. Harris would take everybody and put 'em in the sandhills and hide 'em. Man, he'd clean everything out of here 'cause he didn't want Mr. Forbes to see anybody that wasn't workin'."

E. W. Woody, who retired in July, 1972, as superintendent of the San Juan Division, also has a vivid memory of A. L. Forbes. "Jack Stricklin and I were working down at El Paso City Gate. There were about three or four of us working this gang they had down there, but about forty men showed up every morning looking for work, and you were afraid not to come to work unless you were pretty sick, because if you didn't they might have somebody else in your spot the next day. So it made you think a lot, with all those guys waiting to get your job.

"Mr. Forbes was liable to be any place on the pipeline at any time; anywhere from El Paso to Jal, on to Douglas. You didn't stand around 'cause he might drive up. He drove up on Stricklin and me one time at City Gate. Stricklin was chief dispatcher, and I was supposed to be dispatchin' at the time. We were both sittin' there listening to a ball game and Forbes called in from Salt Flat. Well, we knew where he was so we weren't worried about him. He always drove a Lincoln, and he was one of those people that drove cars like they fly airplanes today, and in an hour's time he walked in the back door. We were both sittin' there listenin' to the radio, and we didn't think he was halfway to El Paso. Forbes was everywhere, out in the field most of the time. He was general superintendent and just naturally every place. That little incident seems funny now, but at the time it wasn't a damn bit funny. He was a rough character. But you had to have respect for him; he was tough but he was real fair. In those days, if it hadn't been for Forbes I doubt we'd ever made it anyway."

As business began to pick up, greater amounts of gas were required from the Jal area and more people were needed to man the mainline compressor stations. R. W. Harris was sent to Station No. 1 near the New Mexico-Texas line to help Estin Scearce who was the station superintendent. Harris recalled going over to help run the old type nineteen engines. "I'll never forget," he said, "Alice and I just put our clothes in a suitcase and went in our own car. We stayed in the old bunkhouse and slept on single beds and had an orange crate and a hot plate and a few pans for our cooking.

"Estin Scearce and a fellow by the name of Mark Roberts would work in the daytime, and Joe Woody and I worked nights.

JOHNNY SCHAFFER, standing at left, was spread superintendent on the first westward extension from El Paso. This picture was taken in 1931 at Anthony Pass in the Franklin Mountains north of El Paso.

This was the whole crew. When Joe and I would get off in the morning we would go hunting, and we killed some of the biggest rattlesnakes you ever saw. Some of them would be as much as four inches in diameter and four or five feet long. We also explored caves that were all back in that area, similar to Carlsbad Caverns. We took flashlights and binding twine and crawled all back in these caves. We strung out this twine so that we could find our way back. Of course, they were nothing as big as Carlsbad Caverns.

"Every two weeks Alice and I would go to town. The town of Carlsbad. The trip was so dusty that we would just put on old clothes and seal our good shoes and clothes in a bag and go to Carlsbad and rent a room. Then we'd go in and take a bath and do our marketing and come back the next day. We got our mail once a week from the pipeline rider. He checked the line and brought all the commodities and supplies back in those days.

"Later Alice and I packed up our tin pans and moved on out to Station No. 2, about a hundred miles away. I stayed and operated it for about a year and then went back to Jal where they made me field superintendent. We really expanded in about 1936. C.L. Perkins moved to El Paso City Gate to take over the job held by C.O. Byrne who had left the Company. H.F. Steen moved from El Paso to Jal to replace Perkins.

"We were putting in some gathering

lines around Jal and Perk decided we needed a tractor. He bought us a T-K 40 International. When the thing came in we were laying some ten-inch pipe. I got our garage mechanic and we went down there and unloaded this tractor and took it out on the pipeline, but nobody could run it. We could do better with a tripod and one of those big jacks. So we just left the tractor sitting back there.

"Well, one day Perk came down there and said, 'Where is your tractor?' I told him nobody could run the thing, and he said, 'Listen you get that damn thing up here and teach somebody to run it!' So we brought it up along the road. An ol' boy came along with his jacket hanging over his back and wanted a job. So I said, 'Say, can you run that tractor?' He said, 'Yeah, I can run it.'

"Well, sure enough he couldn't run it, and when I asked him about it he said 'Yeah, I know, but I need a job.' He put his jacket back on his shoulder and went back trudging down the road. We finally hired a fellow and put him out there with a helper and a joint of pipe and told him to stay on that tractor and stay out of the way so that he wouldn't drop the boom on anyone until he learned how to run it. Well, he did learn how and he got to be an expert side-boom operator, but it took ten days. Perk was very anxious for us to use this tractor because he had spent a lot of money on it. This was the beginning of revolutionizing our way of laying pipe."

After the pipeline to Arizona was built, the usual new set of problems arose to plague the people who were trying to keep gas moving from Jal to the West.

"You might say," Hugh Steen said with a grin, "things started off bad and then got worse.

"After we built this line to Arizona we started having freeze-ups in the pipeline. It was because we changed the pressure in the lines from 500 to 750 pounds and this caused more hydrocarbons and water vapor to condense and drop out in the pipeline. I remember one time we cut the top off a pipeline drip and it was just solid ice. [A drip is a reservoir installed in the pipeline that allows condensate to drip out and keep the line free for gas].

"Well, we took a big chunk of this ice and threw it on the ground. Somebody struck a match and the thing caught fire. It was just full of hydrocarbons. That's one of the big reasons you need gasoline and dehydration plants on a pipeline. In those days we were having freeze-ups every morning, it was just routine. I'd have to get up early and round up a crew and we'd start looking for the spot where the pipeline was freezing up. We'd have to shut down the nearest compressor station to keep from blowing up the pipeline. The hydrates were forming in the twelve-inch pipe and they'd form solid except for maybe a three-inch hole for the gas to go through. You could hear it squealing a long way off.

"Then we would dig out about a four-foot wide ditch and go down about a foot below the pipe, I'd say about fifty feet long. The horrible thing was, just about the time you got it dug out, the block of ice would shake loose and move about a half mile down the line. Sometimes we'd have to build a fire under the pipeline. I went to every one of those tire companies in El Paso that had old rubber tires. Before long I had rubber tires and kerosene stashed all the way up and down the pipeline. I'd have stashes of these tires and wood and two-by-fours and kerosene, and we'd build a blazing fire under there. And boy, that thing would break loose and you'd think there was an earthquake. It would just shake the ground, and if you hadn't been around one of those deals before, it would scare you to death. I don't know that I'd do it anymore. But that's the way we'd have to break 'em loose. In the winter this would happen every night, and the colder the weather got the more the line would clog with ice. So I went through some miserable times until we got the first dehydration plant built down in Jal.

"In 1932, I had moved to El Paso, but out in Jal things got so bad they put us all on half-time. Well, I decided I'd go back to school and get a little more education. So I left, and for two terms I went to Louisiana State University. At the end of the year I wrote—and this is a peculiar thing, how an incident can possibly change one's entire life. I wrote Perkins and said, school will be out pretty soon and how about working this summer? He wrote me back and said 'come on out.'

SHOALES PLANT about two miles west of Jal, New Mexico treated sour gas until it was retired and Jal Plant No. 1 took over its duties in the early 1930s.

"Well, I went all the way to Jal, and Perkins seemed surprised to see me, but he didn't say anything. So anyway, I went to work and after I had been back for about two weeks, I got a letter that had gone to me in Baton Rouge and had finally worked its way back to me in Jal. It was from Perkins saying that things were pretty rough and maybe I better not come after all. Maybe I should find something to do back in Louisiana. Had I gotten that letter I probably would never have gone back to Jal. But Perkins was gracious enough when I showed up, he put me to work in the gang.

"At that time, El Paso Natural owned the gas distribution system of Jal as well as the system to serve the homes and farms

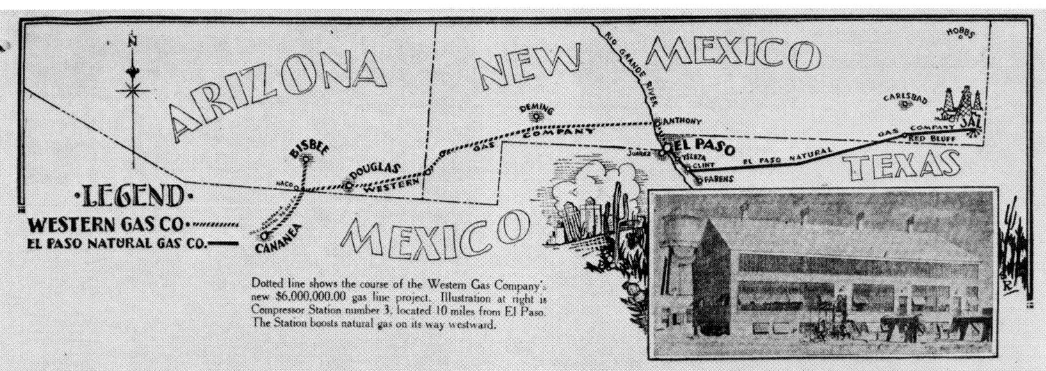

Dotted line shows the course of the Western Gas Company's new $6,000,000.00 gas line project. Illustration at right is Compressor Station number 3, located 10 miles from El Paso. The Station boosts natural gas on its way westward.

WITHIN EXACTLY 2 YEARS

NATURAL GAS WAS BROUGHT TO EL PASO
... AN EXTENSION MADE INTO ARIZONA!

Two years ago, June, 1929, to be exact, the El Paso Natural Gas Company's 214-mile pipe line from the Jal, New Mexico fields to El Paso was completed, and on the same day natural gas was first delivered at the City Gate station for use in El Paso ..An outstanding mark in Southwestern progress was recorded.

Today, the Western Gas Company announces completion of its 290 mile pipe line from El Paso to Southeastern Arizona, and the delivery of natural gas to that section...Another prominent industrial achievement of the Southwest is made.

Progress is dependent upon industry, and industry, for one thing is dependent upon adequate fuel power...The Southwest and its neighbor, Mexico, now have natural gas fuel facilities along an approximate distance of 500 miles...Affording an ideal advantage to stimulate industrial activity, besides enhancing the comforts and conveniences of domestic life.

All around us may be seen signs of progress made in the Southwest during comparatively recent years. . .The Western Gas Company's new $6,000,000.00 project is now another step for this section's ladder of development.

The Western Gas Company has delivered natural gas to its new industrial customers. . . From the viewpoint of purpose of construction all activity could be considered through. . .. But, from the standpoint of our desires and adopted responsibility we will never be through. . . . The intangible items of consistent vigilance and gratifying cooperation in furthering the progress of the Southwest will always be active attributes of these companies.

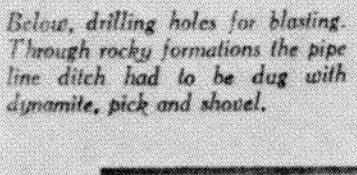

Below, drilling holes for blasting. Through rocky formations the pipe line ditch had to be dug with dynamite, pick and shovel.

EL PASO NATURAL GAS CO.
WESTERN GAS COMPANY
THE PIPE LINE COMPANIES

BASSETT TOWER   EL PASO, TEXAS

FULL PAGE newspaper advertisements announced completion of the new pipeline to southern Arizona.

MECHANIZATION helped speed construction of pipeline to Arizona. By 1931
El Paso's system totaled approximately 500 miles of mainline.

in Fabens, Clint and Ysleta in El Paso's lower valley. They were having trouble with somebody and Clarence Byrne said to Perkins, 'Why don't you send old Steen back to El Paso—maybe he can get this thing straightened out.' So they sent me back to El Paso to run that gas distribution system. Of course, I didn't know anything about gas distribution, but there wasn't anything particularly mysterious about it.

"We had a super salesman by the name of Joe Heinz, and he was just like the Heinz Sauce—a fifty-seven variety mixture of a guy. He could sell a fifty gallon water heater to a one room shack. He was a selling dude! He sold the appliances and I was having to install them. One day Heinz sold a floor furnace to a guy that had a house on a cement slab. The idea was unheard of, but I figured if he was good enough to sell it, I was just stubborn

enough to put it in. With sledge hammers and chisels we dug out a hole in three feet of concrete. I dug a tunnel from outside, ran the gas in there and set in the floor furnace. Well, it worked fine until it rained. The first time it rained, I got a call and went over there. And, of course, the water had run down into his furnace. I had to build a dike around it, but that's the first time I found that gas wouldn't burn under water.

"I guess my first real promotion came along a few years later—in 1938. I got married in 1937, and they sent me back to Jal in '38 as superintendent of the Jal Division. I stayed there 'till 1945 when I came back to El Paso and R. W. Harris replaced me at Jal. They made me superintendent of all pipelines and I worked for Perkins who was manager of pipeline operations."

99

5

THE PIPELINERS

A special breed

By THE LATE THIRTIES, El Paso's pipeliners had come a long way from the time-worn methods they used to lay pipelines only a decade before. Tractors and trucks and heavy earth-moving equipment replaced the horse and mule teams of the Twenties. These men literally pioneered new methods of welding, ditching, crossing rivers, mountains and deserts with hundreds of miles of steel across the toughest country in the Southwest.

Rough, strong rawhide types, this special breed of men known as pipeliners, projected the same indomitable characteristics as the trail drovers that preceded them by a half-century. For the most part, they were unmarried roustabouts who worked hard, played hard, and blew their weekly wages on Saturday nights in the cheap little bars and cafes in the nearest town. The confinement of city life was not for them. They were free-wheeling, independent outdoorsmen whose very existence depended on brute strength, each other, and a sense of humor unmatched in most occupations. This special brand of humor sprang from hard work, low wages, and unbelievably harsh conditions. The pipeliners either had to laugh about it or lose their sanity; living proof that a sense of deprivation can be a blessing for those who can take it.

In short, they were a very salty crew.

One of the saltiest of these pipeliners is a giant of a man named Henry A. Schulze. "Hank" Schulze became a legend because of his exploits, rock-hard independence, and his wonderfully descriptive language. Mellowed somewhat through the years, Schulze finally married an attractive girl named Susan and settled down to a more sedentary life in Farmington, New Mexico as head of El Paso's field operations in the San Juan Division. At six-feet-four and 230 pounds, Schulze is not the type you walk up to on the beach and kick sand in his face.

Hank still speaks with the heavy grits-and-gravy East Texas accent that he brought with him to the pipeline.

Like the story about his pet snake. "Back in 1939," says Hank, "we started the survey on the line from Gage, New Mexico to Miami, Arizona. We had a five-man survey party under Johnny Schaffer's supervision. Chick Walker was our party chief, G. George was our head chainman, and the rest of us just punched stakes and cut brush. This particular crew operated out of Deming, Lordsburg and Duncan, and we always ate dinner at Chicken Smith's Cafe in Lordsburg.

"About that time I caught a bullsnake, and he must have been about five feet long. We kept that ol' bullsnake around, and we'd feed him this or that—got him eggs or rats or somethin'. On this particular day, I had just stuck this ol' snake in my shirt and he wrapped around me 'cause it was cold outside. Well, we went into the cafe and set down there, and the special for the day was roast beef. So we were all settin' there and ordered dinner, and this waitress was comin' down the counter with our dinner and this damn snake got to movin'. I guess he was just wantin' a little air or somethin', but the waitress got up pretty close to the counter and that damn snake started out.

He poked his head out and was just lookin' around, is all.

"Well, that ol' gal was about three or four feet from me before she saw that snake, and then she threw roast beef and gravy all over that cafe! She tossed me out and told me never to come back there again. And I never could eat in there while she was workin'. I had to move down the street a couple or three doors and eat with that ol' Greek. 'Course he put out real good food, but it wasn't as good as Chicken Smith's. I'd go back to get a cup of coffee and she'd say, 'Hell no! You're not gittin', I'm not servin'.' And by god she wouldn't either."

Schulze first went out to Arizona in 1937 to work for a bridge company. "We built a high bridge for El Paso down on the San Pedro River on the old twelve-inch line from Douglas to Tucson," says Schulze. "El Paso pipe crews came in and welded up the pipe, Johnny Schaffer and Pete Delaney and that gang. They didn't have anybody that could get up off the ground more than ten or twelve feet, so they hired a kid by the name of Blake and me to put the pipe up on the bridge. So we did 'em a pretty good job. They thought so, anyhow.

"Schaffer offered me a job, but I couldn't quit this old man I was workin' for on the bridge. I told Schaffer when we got through I'd come and see him. When we finished the job, I went to Tucson and caught 'em all in the Mint Bar on Saturday night and hired out on Monday mornin' for the 'gypsy

"That ol' gal was about three or four feet from me before she saw that snake, and then she threw roast beef and gravy all over that cafe!"

gang' in Arizona. We had five men and a truck and a water barrel and five bed rolls and our grocery box and our shovels and picks and sledge hammers, and that was it. Oh, it was a fine crew and we were gettin' fifty cents an hour. I was only gettin' forty-five on the bridge job, so I made a profit in the change.

"They told us we were comin' out to put a bridge on the San Pedro River near Benson and that, hell, it'd be something like the Grand Canyon. But when we got there it was a little ol' dry wash about a thousand feet across. But it turned out to be a beautiful bridge and *still* it's pretty. If you like bridges, and I did at that time. I thought they were real nice.

"In about '38 they gave me a bossin' job, the first one I'd had with the Gas Company. And the only reason they gave it to me, I guess, was they needed a lot of brute strength. I had a ton of labor to sell and they paid me fifty-five cents an hour, makin' pipe bends and tyin' it together. And right at that time I didn't know anything about makin' those bends, but we had some old men, Blackie Kessler, Little Man Perry, Slim Kirk and guys like that that had been workin' for the big pipeline companies. But they had either drunk themselves out of a job or were just lookin' for anything, and they hired out when we started this 200-mile pipeline. And actually the knowledge they had about bendin' and layin' pipe in those days was a bunch of it. It was a real sharp crew. Now of course, they would take off on ya' and stay drunk for a day or two, but due to the caliber of men and the nature of their existence, they had to get drunk now and then to live with themselves.

HOW TO BECOME PRESIDENT

ONE HOT SUMMER DAY in 1937, a small leak developed in the 16-inch Jal
to El Paso pipeline near Clint, Texas. At this particular point the line runs through the
water table of the area's irrigated farm land, and the leaking gas was blowing up a
geyser of mud and water. A two-man crew was dispatched to the scene to get it repaired
as quickly as possible, but before clamps could be placed on the leak, the hole
had to be thoroughly cleaned out. The fact that the leak was on the underside of the
pipe was no help, either, and the men had to lie down in the mud to scoop it out.

The muddy-looking individual pictured above is Hugh F. Steen, who later became
president of the El Paso Company. Adding to his discomfort was a temperature of 110
degrees which caused the mud to cake solidly on his back. It formed a perfect
oven, and his skin was promptly burned to a turn. The leak was finally repaired,
but Steen spent the next few days in bed recovering from second degree burns.

"I just leaned on them for a little information and they came right through and said, 'Well, if I was runnin' a job I'd do it this way.' And that's the way we laid the first hundred or so miles of the Globe-Miami pipeline, which was the first big one that I worked on for the Gas Company.

"When I started with this gypsy gang we lived wherever night caught up with us. We'd stop by a store and buy our groceries to last a week or two. We had our own tent, skillets, pots and pans and bedrolls, and we would just set up camp and get supper started. We usually carried one barrel of water and it would last about a week if you didn't use too much of it washin' dishes. We didn't do a lot of bathin' unless we came by a stock tank or somethin' like that. It was all right.

"We just did about the same work that the district maintenance crews do today. But in place of havin' heavy equipment we had shovels and sledge hammers and stuff like that. It was head-down, tail-up and go to diggin'. That was all 'cause there wasn't diggin' equipment then. We didn't have a winch truck either. The only winch truck, I think, was in Jal or El Paso.

"It was about this time I got in a little trouble with Mr. Forbes. I turned over a water truck behind Duncan, Arizona on Ash Peak, and this was a brand new water truck. I turned it over on a hill; it didn't have any baffle plates and that water got to sloshin' and I stepped out of the truck and the water just kept sloshin' and in a minute the truck just laid over. I was standin' outside lookin' at it. Oh, godamighty, that

A. L. Forbes

made Mr. Forbes mad. 'Course he'd get mad but he didn't stay mad long if you buckled up.

"Pete Delaney was runnin' a job at Salt Flat, Texas, and Mr. Forbes told me I'd have to go to Salt Flat and work for Pete. He said, 'I'll put you on a tractor and I don't care if you turn it over; it won't hurt much.'

"But anyhow, I told Mr. Forbes I couldn't go to Salt Flat. He said, 'Well, you can go to Salt Flat or get your goddamn time, I'll tell you that.'

"So I said, 'Well you gonna' send it to me or am I gonna' come after it?' He grabbed the telephone and called the office and said, 'Run Schulze off right now. Give him his time.' Mr. Forbes didn't fool around. When he run ya' off, you was gone right then. If you worked for him and you did everything right, you were in good shape. He was a real good man, but when he chewed your tail out, it stayed chewed out for several days. He didn't want no jibber-jabber.

"After Forbes ran me off, I went to work for an ol' boy putting in some little distribution lines for two or three months in Clifton and Morenci. One day Mr. Perkins came by and said, 'Schulze, you want to go to work for the Gas Company?' I told him, 'Yes sir, I'll go down and quit this job right now.' Perk said, 'No you can't do that. Wait 'til you get through laying this line and then you come to Chandler, Arizona. You come to Chandler and work for R. W.' When Mr. Forbes found out about it, he didn't care. And when Perk hired me back he told me, 'Your insurance is paid up, your vacation and everything goes right on like it was. When we fired you you were gettin' sixty-six cents an hour and when you come back I'll give you seventy-two.' So I got a little raise out of the transaction.

"I saw Mr. Forbes many, many times after that and everything was just rosy and lovely. He'd ask me every now and then if I had turned anything over lately or had any wrecks. 'Course I guess I might have been driving a little reckless, but it didn't seem so to me.

"Another thing I'll never forget was when we were surveying a line near Safford, Arizona. We finished that survey, and then the same crew went back and started layin' the pipeline. There was a prize fightin' promoter came down there one day outta' Safford, and we had a few hands that were extremely tough. We had Riley Matheson who later became all-pro linebacker with the Cleveland Rams. We had Ed Alsup and we had John T. Howey, and we had a mess of half-outlaws and individuals who had fought to survive. But anyhow, this guy asked me if I would be interested in fightin'. He just wanted one guy off the pipeline, a four-round bout in Safford, Arizona.

"I said, 'For how much?' He said, 'Forty dollars.'

"Well, I told him, 'Hell yes, I'll fight Joe Louis for forty dollars!' So I started gettin' ready for this prizefight. I'd get these boxin' gloves and spar with Alsup and Riley Matheson and another guy that had been an ex-pro fighter. Well, these guys were mad at me 'cause I had that good deal, gonna' get that forty dollars. Whitey Hobbs was a welder and he was my trainer and manager. Every time we'd spar they'd just try to tear my damn head off if they could, 'cause if they got me crippled up, then they were gonna' move in and get that forty bucks. So we got all kinda' ready and out come the night of the big prizefight and we were gonna' fight in an ol' theater. Well, Mr. Forbes and Mary Magurn and Ralph Allen and Schaffer and Pete Delaney, and all those hands were there.

"I went downstairs to dress and I passed a room and looked in and here's an ol' boy layin' out on a table and they were rubbin' him down and gettin' him all ready. He had his hands all wrapped, and I'll guarantee you when I looked at him, he looked like he was fifteen-foot tall. God, he was a

"I got me some fightin' britches and tennis shoes and went
in that ring and here that dude was."

big 'un. Course he wasn't, he just looked that big to me. But he had on blue shoes and blue trunks and those hands were wrapped up and he was a pretty thing! He was a CC boy from up in Mulecreek.

"So we had bet a little money. I knew I couldn't last over three rounds, 'cause the only trainin' we'd do would be at night at seven or eight o'clock. Just put on the boxin' gloves and spar a little. So we all made our bets. If the fight ended before the fourth round, then I figured I would be the winner. We got all kinda' takers. I mean everybody. So anyhow, the big night was there and I got me some fightin' britches and tennis shoes and went up in that ring and here that dude was. Man he was dressed up pretty as you ever saw. John T. Howey was gonna' referee the prizefight. John T.

Howey looked like a big cat fish. He was all head and no tail. Shoulders and head was all.

"After we'd had our little talk in the middle of the ring, this boy came chargin' across that ring and he started a long, loopin' right hand and hit me right on the shoulder. I went around his right hand and hit him right between the eyes with my left hand. And he hit flat on his back right in the corner and took nine.

"Well, he got up and we waltzed around a minute and I made a little move to one side and he ducked to the other side and I hit him with another left hand and he took eight. In the next round I knocked him down three times, and John T. Howey stopped the fight.

"So we went down to the bar to get our

money and we had won about $150 on bets. The promoter's name was Toughie West and he said, 'We're gonna' call this fight a default, we're not gonna' pay.' 'Well,' I told him, 'the next fight you're gonna' be in, you can be goddamn sure it's a good fight.' He says, 'What'd you mean *I'm* gonna' be in?' And I said, 'Me and you! We're fixin' to have a real good one right here.' And he says, 'Oh no, not you, not you, we're gonna' *pay* you.' I said, 'Well, to hell with that other guy, all I'm interested in is that forty dollars and you'd better come up with it.' 'Oh yes sir,' he says, 'we're gonna' pay you right now.' So I got the forty dollars. Man, what a fight that was."

Along with the battle to build new pipelines throughout southern Arizona and New Mexico, El Paso Natural's whole structure was undergoing rapid changes. In 1937, the Company's first gasoline plant at Jal was put on-stream, and by 1938 it was producing about 12,000 gallons of gasoline a day. The gasoline was stored in nine 25,000-gallon storage tanks and pumped from there to a railroad loading rack.

A year later, the Company was delivering almost 100 million cubic feet of natural gas a day, a far cry from the five million cubic feet deliveries into the city of El Paso just ten years before.

John Eichelmann became chief engineer, with headquarters in the Bassett Tower in El Paso and the department expanded to four people: Hector Valencia, W. A. Mullane and Tom Kerley. The accounting de-

partment had grown to fourteen members.

R. W. Harris was field foreman, and Gus Kendrick was Jal chief dispatcher, C. L. Perkins was transmission superintendent, H. F. Steen was superintendent of what was then called the Jal Division. D. H. Tucker was superintendent of Measurement, and Jack Stricklin was safety director.

Stricklin recalls that the Company almost had its insurance canceled because of the high accident rate, and that's how he inherited the job of safety director. "Mr. Forbes called me one day," says Jack, "and said, 'What are you doin' right now?' I said, 'Right at this minute, nothin'.' And Forbes said, 'Okay, you're goin' to be in charge of safety so we can get some insurance.'

"Things were so bad they couldn't get any worse, so I made out pretty good," says Stricklin. "Since we didn't have a safety man before that, I was the whole schmear. We started havin' safety meetings and I bought books and read up on it and I don't remember any of it, but a lot of funny things happened in that mess. I didn't know if there was goin' to be any more money in it and I was afraid to ask. You know, they'd say, 'Well, its a job isn't it? If you don't like the job, say so.' Needless to say, I liked the job, and they did give me a little raise.

"Later on I taught a first aid course. I was out at No. 2 Compressor Station teachin' a first aid class and Johnny Crutcher came runnin' down there hollerin'

109

that one of the ladies at the camp was ready to have a baby. I just assumed somebody'd do something about it, and I went ahead and turned a page in my book, you know, to go on with the class.

"John said, 'Come on, Jack, this lady's gonna' have a baby.' I said, 'Well, hell, I don't know anything about babies.' He said, 'Listen, you're the first aid man aren't you? If anybody ever needed first aid, she does.' Well, I had a stethoscope that we used to play around with just for fun. We'd listen to each other's hearts, an' hell, half the time you couldn't hear a thing. Far as I was concerned, half the class was dead. As it turned out, they were, from the neck up.

"But I said, 'Well, I'll go over there, but I don't know anything about this. Like all I know is you boil a lot of hot water, that's all I know.' We went over there and this lady was in labor, but she wasn't so bad, and I took out the stethoscope and I made noises and grunts and everything. Reassure her, that's all I was tryin' to do. And that poor soul, she thought she was in capable hands, and if there ever was a novice in the situation, she was lookin' at him.

"Anyway, I got her in a car and we started to town with her, and she'd say, 'I think I'm goin' to have this baby.' And boy, oh boy, that's the last thing I wanted, so I talked her out of it I guess. Anyway, she didn't have it 'til we got to town. But she thought I was just about the best doctor she ever saw. She knew I wasn't a doctor, but she sure thought I knew a lot about babies. That's the closest I ever came to blowing a situation like that and it was the last time.

After that, when somebody'd say a lady was goin' to have a baby, I'd say good for her, we need the population. That's as far as I'd ever go."

In 1939, the Company celebrated its tenth anniversary with ten-year service awards to Paul Kayser, Miss A. C. Martch, A. L. Forbes, V. V. Vincent, C. L. Perkins, J. G. Bennett, H. J. Clifford, L. M. Stricklin, Leo J. Weske, J. E. Franey and E. E. Walsh. The Company magazine, *The Pipeliner,* which had begun publication a year earlier, with its cover designed by Hector Valencia, published a special feature on the anniversary. Paul Kayser, speaking to these first ten-year employees had this to say: "At this time, when almost the whole world is engaged in one of the bitterest struggles of its history involving essentially only economic problems, it should be a matter of greatest satisfaction to you and to all of us that we have been able, out here in this desert, to solve our own individual economic problems with such a measure of prosperity and happiness as we have been able to solve them here. You are to be greatly congratulated upon the fact that you have had the hardiness to come out here in this generally desolate, sandy country and stick to your job until you have now transformed it into a fairly decent place in which to live, and at the same time to build this Company into the profitable enterprise that it is."

Paul Kayser's speeches to employees were always enthusiastic, positive and always off the cuff. No notes, no speech writers. He believed in his people and his sincerity filtered down to all levels regardless of rank.

From the very beginnings, employees were never just numbers and names on personnel records. To this day, his compassionate influence remains strong on all employees—from the boys from Clyde and the Depression days, to the younger groups of men and women taking their own places in El Paso's operations in the Seventies.

Typical of this feeling was an incident that happened to Hank Schulze back before the war. "We were out layin' this little ol' ten and three-quarter-inch loop line out of Tucson back toward Station No. 6 and my stomach got to hurtin' so that I couldn't pick my feet up on this tractor," says Hank, "and at that time they didn't have much in the way of fringe benefits. I was hurtin' so bad I told ol' Pete Delaney, 'I've got to go to town now, I'm hurtin' too bad.' He said, 'By god you can't go now, I ain't got no other tractor driver. You'll be alright this evenin'.'

"Well, in a few minutes here comes ol' Paul Walsh, the timekeeper, and I just dropped off that tractor and got in his pickup. I told him, 'Take me to town.' So he wheeled me into the doctor's office. We just went down the street 'til we saw a doctor's sign and pulled in there and I told that ol' man, 'I've got a stomach ache so bad I can't even move.' He said, 'Have you got any transportation? Can you get to the hospital?' I said, 'Yes sir.' Well, we got to the hospital at 12:30, and hell, at 1:30 they had me shaved and shined and in there on the operatin' table, and they took my appendix out.

"Well, lord, when I got out of the hos-pital, Mr. Forbes came by and said, 'Alright now, you've got a week's vacation and a week's sick leave and when them two weeks are over you oughta' be able to go.'

"Right after that I was back in Jal and I got word my wife was real sick in Waco, Texas. I had to go down there and got a special nurse and drove her from Waco to Phoenix, Arizona. Then I brought this special nurse back to Monahans and put her on the bus for Waco. All that time and all the money I spent, I didn't know how I was gonna' make it. I drove all the way from Phoenix, let that nurse out at Monahans about four or five o'clock in the mornin' and drove to Jal. I put my work clothes on and went to work at seven.

"But anyway, when I got to Jal I went in there to the office and Mr. Steen asked me, 'Well, how did everything come out?' I said, 'Real good, everything's real good.' So I went back to work and I worked to the end of the week. I knew when the pay checks came in I figured I was goin' to have maybe twelve or fifteen dollars, something like that. But here comes my whole paycheck, right on through. Godalmighty, I'll tell you what—nobody knows how much a hand appreciates stuff like that when he's really in a bind. You know, it gives you somethin', because I'd spent a lot of money on gasoline and all the other and I was flat broke. I really didn't know what sick leave was or how it was workin'. Hell, it just came through. It sure made Mr. Steen a big man in my eyes, I'll tell you that."

Nationally, people were beginning to lose their Depression blues, even though WPA workers were still leaning on shovels all over the country. Major Bowes' Amateur Hour was the hit of radio. A freshman at Harvard gulped down a goldfish on a bet and started a fad which didn't end until someone downed forty-two in succession. Shirley Temple was the nation's darling, and the country mourned when Will Rogers was killed in an airplane crash with Wiley Post at Point Barrow, Alaska.

Millions of kids sat glued to radios listening to Jack Armstrong, The All-American Boy, win football games for dear old Hudson High while the sponsors sold tons of boxes of Wheaties to parents throughout the land. Equally exciting were the radio adventures of Little Orphan Annie, complete with premiums like secret decoder rings and shake-up mugs for chocolate flavored Ovaltine.

Teenagers and their elders were singing such classics as "The Music Goes Round and Round," "A-Tisket A-Tasket," "Jeepers Creepers," "Three Little Fishes," and "Ti-Pi-Tin." They just don't write songs like that anymore.

But war nerves began to take over from the blues of the Depression. German troops marched into Austria in 1938, and Prime Minister Chamberlain with his umbrella made the famous trip to Munich in a futile attempt to pacify Hitler. The world sat up late at night to hear H. V. Kaltenborn chronicle the world on the brink of war. Chamberlain returned to London proclaiming there would be "peace in our time." But a trio of egomaniacs named Hitler,

JACKHAMMERS were used to cut through rocky pipeline right of way during extension from Red Rock to Hayden, Arizona in 1937.

112

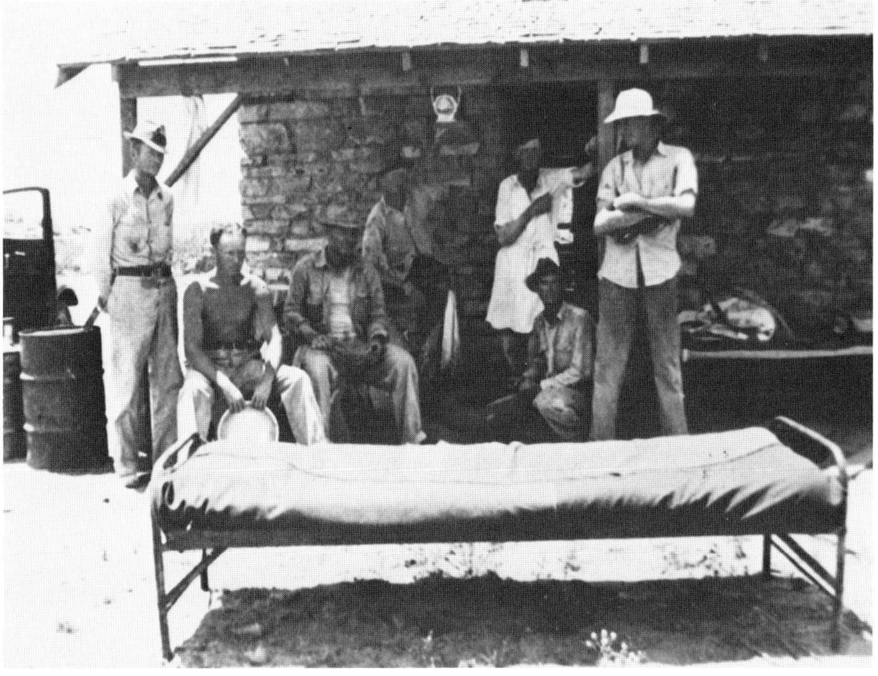

**SMALL
DIAMETER**
pipeline spans a dry
wash in saguaro
forest from Casa
Grande to Superior,
Arizona in 1936.

**ALL THE
COMFORTS** of
home are enjoyed by
these pipeliners in
West Texas in 1936.
They worked and
lived under harsh
conditions wherever
construction was
going on.

113

Mussolini and Tojo were soon to change all that.

The atomic age began with the inauspicious announcement that two German scientists had discovered uranium was fissionable, but the news was crowded off the front pages by other, more easily grasped events. Germany invaded Poland on September 1, 1939, and Great Britian and France declared war on Germany. Italy invaded Albania, and Russia invaded Finland.

The war still seemed far away to most Americans in 1939.

In 1940, Wendell Willkie was the great hope of the G.O.P.—at least part of it—and Franklin Roosevelt confidently planned another series of fireside chats. The armies of Great Britian and Germany faced each other across the Channel, and a man in Pittsburgh offered a million dollars to anyone who would kidnap Hitler. In Washington, Colonel Charles A. Lindbergh predicted the war would end in a stalemate and suggested a negotiated peace.

Also, in 1940, El Paso Natural Gas pioneered one of the greatest conservation efforts of a natural resource in the country's history when it began marketing "residue" gas from fields in West Texas and New Mexico. This gas had formerly been burned in huge flares because of the lack of a market, and Jal Plant No. 2 was constructed to handle this residue gas. Residue gas is derived from "casinghead" gas which is produced in conjunction with the production of oil. Natural gas, contained in oil, escapes in much the same way carbo-nated gas bubbles out of a bottle of soda pop when the cap is removed. Natural gas is separated from the oil in individual separators at the field, gathered to a central point and processed for gasoline, propane, butane and other hydrocarbons. The natural gas then remaining is called residue gas.

From the 1920s until the mid-1940s, every oil wellhead had a flare. Long-time employees from the Jal area remember the countryside being lit up for miles with the eerie, yellowish glow from the oil well flares. The distance between Carlsbad and Hobbs is about a hundred miles, and Paul Trout recalls that "from the Halfway Bar you could see the reflection of the Jal flares at night. One huge flare at Jal No. 2 was burning about seventy or eighty million cubic feet of gas a day. Man, it would be nice to have that now," he says.

J. C. Williams of Jal says that "you could stand up here on Phillips Hill [a little hill right north of Jal] and look south and it was just like daylight out through there, flares burning everywhere twenty-four hours a day."

He also remembers driving around at night in the Goldsmith area before El Paso put in its Goldsmith Plant. "You could drive all over the place without even using your headlights. People even built houses out there at night without even using additional lights."

The question arises as to who got the idea to put out the flares; did it just evolve or how did it come about? According to John Eichelmann, "All that credit is given to Mr. Kayser. He saw all that gas being flared and wasted, and it was a spectacular thing to see around Jal. At nighttime it would illuminate the area all the way from Hobbs down to Jal. Mr. Kayser had the vision to build a plant and get this wasted gas into our pipeline system."

As the construction of Jal Plant No. 2 got started under the supervision of C. C. Selley, the town of Jal began to boom all over again as workers and money poured in. Previously, it had been a typical oil town, witnessing the incessant monotony of the arrival and departure of many people. But with the growth of El Paso Natural Gas, a more permanent upsurge of activity filled the town. Everyone felt the monetary effect of the building of the new plant.

John Eichelmann says the toughest job he ever tackled for the Gas Company was the construction of this plant. "We were pioneering something that I didn't know anything about," says Eichelmann, "and very few others knew anything about building a plant like this either."

El Paso had made a contract with Phillips Petroleum Company in December, 1939, to take casinghead gas from them and have a plant completed and ready to process the gas by April 15, 1940. El Paso's construction crews broke ground on January 26, 1940 and almost miraculously the plant was in operation and producing gasoline on the night of April 14, ahead of the deadline. The Company did all its own engineering with its small staff in El Paso. They ground out drawings without a break to keep ahead of the construction crews.

IN 1937 El Paso's pipeliners laid an extension through desert from Red Rock to Hayden, Arizona.

WIDE BRIMMED HATS brought a measure of relief from searing West Texas sun near the Pecos River in 1936. Cots in background made for gracious, cool sleeping outside at night. Most pipeliners acquired nicknames as evidenced by this group. From left to right: "Fats" Pridgeon, "Speedy" Stocks, "Flop" Dobbs, Joe Walden, "Pud" Sapp and Conard Cotton.

Blueprints made during the day were rushed to the post office so that C. C. Selley could pick them up the next morning in Monahans and keep the job moving without a let up. Up to that time, according to Eichelmann, there was no other plant quite that large; this one could process eighty million cubic feet of gas a day, and the related compressor building itself was 314 feet long. "The complex made Jal Plant No. 2 the largest gas absorption plant of its kind in the world at that time," says Eichelmann.

By 1940, the number of El Paso's regular operating employees had reached a record high of 413.

In 1941, the first peacetime draft in United States history was well under way, and El Paso Natural lost its share of people to the service. In letters to departing reserve officers and draftees, who had just received their greeting from Uncle Sam, General Manager C. C. Cragin promised the departing employees that El Paso would be looking forward to their return, and released the welcome news that the Company would pay their group insurance premiums during their absence. As the government began building facilities at Fort Bliss in El Paso, crews laid pipe to the "cantonment," as it was called, for heating and cooking.

Interspersed with draft and reserve officer news in *The Pipeliner* were items about employees in the National Guard called to active duty, some of whom would, in just a few short months, be making the Bataan Death March to Japanese prison camps.

One employee's wife made these remarks in *The Pipeliner* in a story titled *What it Means to be a Gas Company Employee's Wife*: "What do we get when you get home? We get a cheerful 'hello,' some Btu's, a stack of charts, great big engines

SMALL pipeline extensions throughout southern Arizona continued through
the late 1930s. By 1940, the number of El Paso's regular operating employees
reached a record high of 413.

and so many thousands of feet of pipeline.
Do we understand it? Well, no we don't, or
maybe yes, vaguely. But we are interested—
definitely. We understand that you like
your work and that's why we get the cheerful 'hello' in the first place, so bring your
valves, calorimeters and blowtorches. We
love them."

With the completion of survey work for
pipeline "loops" (parallel lines) to be laid
from Clint to Station No. 3, Salt Flat
toward Station No. 2, from the Guadalupe
Rim Rock toward Salt Flat, and between
Jal and Station No. 1, the last gaps in the
sixteen-inch single pipe mainline between
Jal and El Paso were eliminated. Looping
was also done on the Tucson-Phoenix line
and the Company extended service to the
military bases which were beginning to be
built and expanded in the area. Compressor Station No. 7 was constructed near
Morenci, Arizona, and by late 1941 system

capacity had increased to 130 million cubic
feet of gas a day.

While war clouds were gathering, most
Americans assured themselves that two big
oceans separated them from the quarrels of
foreigners. They preferred instead, the innocence of not worrying about the future
and continued having fun dancing to the
big band swing music of Benny Goodman,
Artie Shaw, Glenn Miller, Tommy Dorsey,
Count Basie and Duke Ellington whose
hard-driving beats and improvised solos
bellowed from Wurlitzer juke boxes
throughout the country. Featured vocalists
like the Andrew Sisters, Ella Fitzgerald and
Martha Tilton helped put across hit tunes
such as "Playmates," "I'll Never Smile
Again," "Sunrise Serenade" and countless
others.

On December 7, the United States was
jolted out of its customary Sunday torpor
to find that it was at war—and doing badly.

AS WAR PRODUCTION
in the U.S. moved into
high gear, El Paso Natural
quickly accepted the
challenge to provide energy.

6

THE WAR YEARS

AT 7:55 A.M. on Sunday, December 7, 1941, America lost its innocence. The first wave of more than 350 carrier-based Japanese planes appeared out of the mist at Pearl Harbor to rain their deadly strings of bombs and torpedos on the American fleet which bobbed lazily at anchor. In slightly less than two hours the Japanese attackers sank or damaged eight battleships, three light cruisers, three destroyers and four other vessels. At Hickam Field, the attack destroyed more than 170 planes; 2,400 Americans died that Sunday morning.

Although the attack on Pearl Harbor completely surprised most Americans, Japan and the United States had been drifting toward this major confrontation for at least a decade. A collision course became almost inevitable with the increased rivalry between the two countries for supremacy in the Pacific; Japan invaded China, and announced plans to expel whites from Asia.

The United States, in order to stop further development of Japan's war machine, completely halted all shipments of oil and scrap metals to the Japanese in 1940. Tensions between the two countries increased when an extremist Japanese military cabinet took office, under the leadership of General Hideki Tojo. Actually, Japan's militarism asserted itself as early as 1931 with the invasion of Manchuria, and some historians consider that was the real start of World War II. In 1940, Roosevelt froze all Japanese assets in the U.S. A weakened Japan felt that the only alternative was to strike the United States while this country was still pitifully unprepared to wage a major war. And this they did on December 7.

The U.S. had every possible warning, particularly since the military broke Japan's top diplomatic code almost a year before. Yet when the attack came, Americans were stunned to find themselves unbelievely unready for battle. "Retreating to new defensive positions" became a byword in the U.S. press. On December 11, Germany and Italy declared war on the United States, and Congress reciprocated.

America experienced one grim defeat after another, particularly the complete surrender of the Phillippines. Then in late 1942 the country began to stem the Japanese tide of victory.

In the meantime, as war production in the U.S. moved into high gear, El Paso Natural Gas quickly accepted the challenge to provide energy for the country's military production. C.F. "Preach" Rittmann, editor of *The Pipeliner*, said to employees in March, 1942: "During the past month the reports coming from the battlefront have been anything but encouraging. These reverses of our armed forces have been attributed to lack of equipment both as to quantity and quality in comparison to the Axis powers. Their numerically superior forces and greater amounts of material have made it possible for our enemies to obtain their every objective.

"We hear much quibbling and bickering about this lack of equipment, relative to the blame. But the placing of the blame for this woeful lack of equipment for our armed forces is now academic.

"Are our embattled forces, fighting gallantly, going to receive help by having blame placed on someone's negligent shoulders? Those men in the Philippines, do they want words explaining why they have no equipment or would they not prefer action? Let's forget about the blame, and in its place let's substitute shiploads of tanks and bombers and fighters to darken the skies with American planes and give them a chance to remove the deadly yellow peril that now perches arrogantly from the cockpits of planes wreaking devastation upon them.

"The past few years have clearly demonstrated that our men cannot be provided necessary equipment by wishful thinking. It can be done, though, by a determined effort of the greatest industrial nation in the world. Industry now is geared to its mightiest potential. Our entire nation has accepted the challenge to produce quantities such as the world has never seen.

"As employees of an industry vital in defense efforts, we are a part of the great mass that our soldiers are directing their appeals to. And we have this brought home to us directly by one of our own former employees. Our answer to the appeal, 'Keep the gas flowing and I will help keep 'em flying,' can best be given by the results of our efforts to work such as we have never worked before. And when the hours seem long and the labor tough, think of the boys on Corregidor Island. They have no forty-hour week, no eight-hour day. Their shift is twenty-four hours. They are not complaining, only appealing 'send us supplies.'

Our answer: 'We will not let them down.' "

These words today may sound a bit

melodramatic, like a TV rerun of a wartime John Wayne movie. But in the early part of 1942, words like these had a real message for the folks back home reading the papers and listening to the radio about the distressing retreat of their fellow Americans doing battle with the Japanese.

El Paso Natural's employees responded for duty by the dozens. Like John Sparling, who was one of the first draftees. "We went over to the Jal recreation hall where we had safety meetings and stuff like that," says Sparling, "and they gave us all a number. My number was 368 and when we heard they were going to announce the lottery, we sat around and listened to the radio. And sure enough, the first number I heard was 368. It was so quick, I didn't know what happened to me. I had just bought me a new '40 model Ford for $42.40 a month. I didn't know how I was goin' to pay for that. I owed for everything. I had even bought a new suit from Sears and Roebuck and I owed for that. Seven of us went out on the second draft call from Lea County, New Mexico, and the next thing I knew we got off the troop train in Fort Sill, Oklahoma. In those days they didn't have any new GI issues. It was all World War I stuff. They didn't pay any attention to size either. It was raining cats and dogs and we had to stand at attention out there in this rain. Then some sergeant came by throwing these overcoats at us. I got one that must have been used by an eight-foot boy in World War I 'cause I had to hold it up out of the water And the sleeves came way down past my hands. But anyhow, it kept the rain off."

Another draftee from the Jal area was J. C. Williams who was living in the little shanty town of Bennett when he received his notice. "Bennett wasn't marked on the maps when I went into the service in '41, but it was my mailing address," smiles Williams. "The government held up my mileage and travel allowance for a few weeks because they couldn't find Bennett, New Mexico. Of course, they could have looked it up and found out that Bennett had a post office, but they didn't know where it was. I tried to tell 'em but they wouldn't believe it. I got a lot of kidding, coming from a place that wasn't even on the map."

Letters from the Company's men in the service poured in by the hundreds and were printed in every issue of *The Pipeliner* during the war years. Most letters expressed fond memories of their past jobs and their all-consuming desire to return to their civilian pipeline jobs.

Meanwhile, El Paso's pipeline crews were doing their best to provide critical wartime energy. R. W. Harris remembered one job that the Company handled for the Bureau of Mines. "Perkins took this job to lay a pipeline from Shiprock, New Mexico to Gallup, a distance of about a hundred miles to tie-in some helium wells. There was a helium plant built up there and the government needed the helium for balloons all over England to help snare German bombers.

"We went up there right in the big middle of winter and with the help of

Navajo Indians laid this pipeline," said Harris. "This was our first acquaintance with the Navajos and it became very important in later years when we moved into the San Juan Basin. I had the opportunity under Perk to go up there and do this job. I never will forget, it was getting along toward Christmastime and it was a brutal, cold year. Even during the war they hadn't perfected warm clothes and we were all freezing to death. I called Perk and asked him what he thought about shutting down for Christmas, maybe for a week and then going back; that the men were cold and sick. Actually, they weren't as sick as I was making out. Perk told me, 'Now look, if this is too rough for you, just let me know and I'll send somebody else up there.'

"Well," Harris chuckled, "I went back out there and told these fellows, now look, we have a job to do and we are going to do it. If it is too rough for you, just let me know and I'll get somebody else."

The sky that winter was like wet cement and the wind, straight out of the north, made everybody's ears ache, but Harris and his crew finished the job.

Steen remembers that Harris and his men had to go out and dig new lengths of pipe out of the snow to help meet the time deadline. "Perkins was driving them pretty hard," says Steen. "I'll never forget the story about the old boy from Lousiana. They were all about to freeze to death, even while standing around a fire one evening. This good old Cajun boy said, 'Listen, don't you hear that?' And they said, 'What are you talkin' about?' And he says, 'I hear my mama callin' me and I'm goin' home.' And he just walked off down the road—and you couldn't blame him."

In 1942, El Paso Natural began putting more and more gas through its pipelines as a result of the increased demand. Field employees started scratching their heads figuring seriously how they were going to travel their accustomed 75,000 to 80,000 miles a year with tires and gas in short supply. The Company instituted an extensive program for materials conservation; metal goods, tires, gasoline, everything. As men left to go into the armed services, more and more women, as in all industry, filled out the roster.

With more and more gas being funneled into vital defense industries and military installations throughout the Southwest, El Paso instigated a tight security program to protect its flow of energy. Armed guards appeared at the gates of each compressor station and plant and the special pipeline patrols kept up a continuous watch over the long, lonesome stretches of right of way to keep out any intruders.

On March 10, 1942, employees received the distressing news that H. G. Frost had died after an operation at the early age of fifty-one.

By that time the number of employees had increased to 741, including an accounting department—composed of twenty-three people. All Home Office employees were still quartered in El Paso's Bassett Tower, business records kept in the basement, and

DURING WORLD WAR II El Paso Natural's company magazine,
The Pipeliner, carried its share of patriotic cartoons and illustrations. More and
more women filled out the Company's employee roster.

Ralph Allen and Bill Monahan handled the entire payroll. In November of that year a new plant at Eunice, New Mexico went into operation, compressing gas through a newly-completed pipeline serving Carlsbad and the vital potash industry. This upped system capacity to 146 million cubic feet of gas a day and the Company got a look at a situation that it would see again in the future; the first rate case came up with the Federal Power Commission.

In 1943, El Paso Natural was supplying the fuel requirements for more than forty-two percent of the total new production of vitally essential copper produced in the United States. It also supplied fuel for other war metal production (a magnesium plant and two steel plants), twenty military installations and sixty-two communities in Texas, New Mexico and Arizona. Employees planted Victory gardens and canned their own fruit and vegetables.

A. L. Forbes, who had been so instrumental in molding the Company in its early years, left El Paso in 1944 to join the Pressure Weld Company in Houston and C. L. Perkins took over as general superintendent of El Paso Natural Gas, succeeding Forbes. H. F. Steen, who had been division superintendent at Jal, was promoted to take over Perkins' former spot as assistant general superintendent of the Company. Fred Wagner joined El Paso Natural in 1945 as a Vice President and assistant to the President, Paul Kayser. Wagner was elected to the Board of Directors the following year. R. W. Harris became division superintendent at Jal, and Jack Smith succeeded Harris as field superintendent.

At Jal, things were humming. The construction department remodeled the old Jal office building, whereupon H. F. Steen's newly-decorated office, with three upholstered chairs, became known as "Plush Palace." For the first time, new field lines began probing West Texas in search of more gas supplies.

For more than nine years, Jal had been without passenger train service but the war changed all that. Resumption of passenger service was an occasion for a general turning out of the populace. School was let out, a band played, and two lady passengers who were to be the first to board a train at Jal in almost a decade, were so overcome by the excitement that they forgot to embark, and the train started off without them. The agent had to flag the train, which

124

backed up so that the ladies could be safely deposited aboard as the crowd cheered.

Patriotic Americans, in addition to hearing Edward R. Murrow report the war news on radio, listened misty-eyed as Frank Sinatra warbled the sweet tunes of the early Forties on "Your Hit Parade," and they hummed wartime classics like "I Left My Heart At The Stagedoor Canteen," "Don't Sit Under The Apple Tree," "The White Cliffs of Dover," "Praise The Lord And Pass The Ammunition," "Der Fuehrer's Face," and "When The Lights Go On Again All Over The World." Since it had been only a little more than two decades since World War I, lots of people felt that these ditties didn't compare with the patriotic songs of those days. But Americans by the millions heartily sang the new ones which were filled with nostalgia and a certain amount of corn.

On the war fronts, the dismal American defeats of 1942 began to shift, although a Japanese sub shelled a California refinery and a German sub sank a U.S. destroyer off Cape May, New Jersey. Major Doolittle's daring raid on Tokyo was a tremendous morale booster, as was the flash, late in 1942, of the U.S. landing in North Africa under the leadership of General Dwight Eiesnhower. People would have cheered if they could have been told of another top-secret operation: the first nuclear chain reaction, produced at the University of Chicago, which laid the groundwork for the development of a weapon that would end the war—at Hiroshima.

American troops began to tighten a noose around Japan by launching more than 100 island invasions from one obscure Pacific Island to another: Guadalcanal, Tarawa, Eniwetok, Saipan, Iwo Jima, Okinawa and countless almost unpronounceable others.

In Europe, Italy surrendered, and just to show how things can change, a German bomber purposely sank Italy's 35,000-ton battleship *Roma* as it steamed toward an allied port. In Washington, Secretary of State Cordell Hull branded statements by a capital columnist that he was anti-Russian as "monstrous, diabolical falsehoods."

History's greatest armada landed history's greatest invasion force in Normandy, and after breaking out of the French hedgerows, G.I.'s chased the German Wehrmacht clear across France until — it could only happen to an American army — they ran out of gas. In 1944, with corn-cob pipe firmly clenched in his teeth, General Douglas MacArthur waded purposefully ashore in the Philippines.

President Roosevelt died, and Vice President Harry S. Truman assumed the leadership of the country. Germany capitulated in 1945, with Japan following after the A-bomb drops at Hiroshima and Nagasaki. The best estimates show that World War II took the lives of fifty-five million people.

With the end of the war, businessmen were free once more to go after the markets and pent-up demand which years of wartime restrictions had built-up.

At El Paso Natural Gas Company, the stage was set for spectacular expansion.

7

TO CALIFORNIA

The flares go out

DURING WORLD WAR II California exploded into a boom the likes of which had never been experienced in history. Sprawling southern California aircraft factories supplied thousands of fighter planes and bombers that helped turn early defeat into victory by 1945. Its shipyards, naval bases, steel mills, and countless other industries provided muscle for America's war arsenal. The Okies, Arkies, Texans and workers from all over the country headed for the gentle climes of California and the golden opportunity to benefit from wages of the defense industries. California looked like heaven to them, especially to those who had survived the lean Depression years and the Dust Bowl of the Thirties.

Los Angeles in 1940 was a magic lure for families from the Southwest arriving daily in their wired-together jalopies and Model-A pickups crammed with kids and worn out furniture with the inevitable mattresses strapped on top. These were days before "smog" became a dirty word. Highway engineers built the first freeway, from downtown Los Angeles to Pasadena, and in wintertime, the majestic, always visible snow-clad Los Angeles Mountains ringed the green rolling hills and citrus groves of the basin.

The wartime influx of population, along with defense plants grinding out massive quantities of war materials twenty-four hours a day, placed a heavy drain on its reserves of oil and natural gas.

But the boom of the war years soon

turned out to be only the tip of the iceberg. The post-war boom was incredible. Thousands of ex-dirt farmers, with vivid memories of the drought of the late Thirties, decided to remain on the West Coast. Servicemen and women, mustered out in 1945 and '46, liked what they saw in California and they, too, decided that California was indeed the promised land. Route 66 and Highway 80 sagged bumper to bumper with a steady stream of new families leaving the past behind to stake out a future in the vibrant Golden State.

Southern California became a national phenomenon. With sprawling Los Angeles at its heart (referred to by eastern pundits as "the biggest hick town in the world"), freeways appeared, and the wide asphalt and concrete ribbons snaked out in every direction to ease the traffic on choked streets leading to new subdivisions. Los Angeles had more bottlenecks than a distillery as mushrooming clusters of homes appeared almost overnight in cow pastures, bean fields and orange groves. One new city, Lakewood, grew larger than Lexington, Kentucky in less than two years. Within a decade, after the end of the war, the San Fernando Valley held more people than Minneapolis or Buffalo. Every year Los Angeles alone added more people than the population equivalent of Salt Lake City, Utah at that time.

There had been nothing like this explosion of growth since the early flood of European immigration built New York City.

The amount of energy required to fuel this kind of growth was undreamed of, and it became apparent all too soon that California was literally running out of gas. The state had to look elsewhere for its energy supply, and the nearest source was Texas. El Paso Natural Gas Company, with its already existing pipelines running into western Arizona, was in the right place at the right time.

Actually, the first thought of a project to bring Texas gas to California occurred to Paul Kayser and A. L. Forbes as early as 1938. They reasoned that by doing a little "looping" on El Paso's existing eight-inch line to Ajo, Arizona and building another hundred miles of new line to Yuma, they could cross the Colorado River into California and deliver fifty million cubic feet of gas a day to the southern California gas companies. Of course, the war intervened and El Paso had its hands full just supplying its own little market in the Southwest.

Kayser recalls that in 1945, "We learned that California was looking for some gas from out of state. So Cragin and I made an appointment with Frank Wade, president of Southern California Gas Company, and we discussed his needs. We told him that we had this line extending to Ajo, Arizona, within a hundred miles of Yuma on the California border, and that we could easily deliver him fifty million cubic feet of gas a day. We understood that was all he wanted.

"But we were wrong. He wanted seventy-five million immediately. Furthermore, he wanted another fifty to 125 million shortly afterwards, and a step-up over two or three years to 300 million cubic feet of gas a day."

Since El Paso's total daily sales in 1945

amounted to only 136 million cubic feet a day, the quantity asked for by Southern California was mind-boggling. It would mean almost tripling the size of the Company. Kayser had heard rumors that Southern California was already talking to several other firms, but the idea of supplying an additional 300 million cubic feet of gas a day was startling. After the meeting with Frank Wade, Kayser says, "I put my stuff back in my briefcase and started to get up. Wade said, "Where are you going? Don't you want to work out a contract here with us?' I said, 'I'm going back to El Paso. And as fast as I can do it I'm going to prepare an application to be filed with the Federal Power Commission to sell you this gas. I'm not going to be the second one to file the application to bring out-of-state gas to California! I'm going to be the first.' "

Phillips Petroleum Company owned large amounts of gas in the Permian Basin as well as in the Hugoton Field of the Texas Panhandle, and El Paso negotiated with Phillips to buy gas from them and move it to California. Meanwhile Kayser flew to Washington, and on August 10, 1945 he filed the application. El Paso's commitment to sell gas to the California customers (at the Colorado River) at the low price of 13 cents per thousand cubic feet was enough to beat any competition from other sources.

A great gas transmission line provided the answer for California, and the increasing demand coincided with the growing needs of Arizona, New Mexico and West Texas. But no ordinary pipeline would serve. It had to be big. A pipeline twenty-six inches in diameter would carry the load so vitally needed throughout the Southwest. Even though construction costs skyrocketed after the war, the building of a pipeline would provide all the additional gas required in this wide area at the same low price as before.

El Paso's application with the Federal Power Commission called for building a twenty-six-inch pipeline to California, running from the gas fields of West Texas and southeastern New Mexico to the Arizona-California boundary near Blythe, California on the Colorado River. Also included, was construction of a twenty-four-inch line to tap the Panhandle and Hugoton fields in the Texas Panhandle at Dumas, moving gas southward and connecting with the twenty-six-inch line in Lea County, New Mexico.

Two of the country's great gas distribution companies, Southern California Gas Company and Southern Counties Gas Company, already served a population of more than four million people but the supply of available gas declined steadily. Under the proposal, they would build their own pipeline and facilities from the Colorado River to the Los Angeles area. Initially, El Paso was to deliver 125 million cubic feet a day of residue gas (produced in conjunction with oil) the first year, and increase the load to 175 million soon thereafter until ultimately the capacity would be increased to 305 million cubic feet of gas a day.

The whole project had a conservation appeal to the FPC because seventy-five percent of this gas destined for California was being flared to the air and wasted in the fields because of a lack of market. The rest of it would come from "dry" gas wells in the Panhandle and Hugoton fields of Texas. El Paso contracted for seventy-five million cubic feet a day from Phillips Petroleum Company, thirty million feet from Gulf Oil Corporation, twenty million from Shell Oil Company and ten million from Warren Petroleum Company.

Hugh Steen remembers very vividly the flaring gas. "There was so much gas flaring in the Permian Basin you wouldn't believe it. It was sour—full of hydrogen sulfide and other impurities—and of no use until it was purified. The gas, mixed with oil as it came out of the wells, was separated —leaving the gas to flare out of a vent line. The more valuable oil was trapped in separators and then piped to storage tanks. One way we sold the California people that we could provide the volumes they needed was to bring them out to Hobbs. We acquired an airplane—a Lockheed Lodestar. And we'd take 'em at night on a ride over that oil field. It looked like just one big flare across the whole field. If they had any doubts about the gas being there, they pretty well lost them. They were also impressed by the tremendous number of wildcat wells being drilled with a ratio of high success. As it turned out, we were the first company that ever took large quantities of residue gas and used it like we did.

"When we decided to build this pipeline to California, well, there can never be enough credit given to Perkins for the many things he came up with. He was extremely innovative. He still is one of the master pipeliners of the world, of the century; there's not any question about that."

Paul Kayser recalls that the Texas Railroad Commission held a meeting to solve the problem of flare gas—a meeting that would have an all-important impact on the Company's plans to move gas to California. "The Commission," says Kayser, "had just shut down the Permian Basin entirely and stopped production of oil until proper disposition could be made of the flare gas.

"All the major companies were represented at this meeting, and the chairman of the Commission asked for a statement from them as to what they were doing to conserve gas. One after another they stated that they had made arrangements with El Paso Natural Gas Company to take this gas into its system. The chairman turned to me and said, 'Mr. Kayser, what does El Paso Natural have to say about that?'

"So I told him that we had agreed to take all flare gas produced in conjunction with the oil in the Permian Basin. This closed the matter, and the Commission lifted its restrictions on oil production.

"This was one of the proudest moments of my career. The action of El Paso led the way and set the precedent for marketing residue gas. Within a few years the Company was selling about a billion cubic feet of gas a day — a substantial contribution to conservation. Without this residue gas, El Paso couldn't have made the tremendous expansion that it did during those years.

"And of course, I couldn't have made the sweeping statement that El Paso would take all the residue gas then being flared in the Basin without knowing beforehand that the Company had already developed, through our engineers and chemists—and our actual experience—methods of treating and handling this gas, an essential and difficult task."

Through the years of trial and error, El Paso gained the necessary experience to tackle a major undertaking like the California line. Its pipeliners had already laid hundreds of miles of main and gathering lines in the Jal field. They also handled contract work for a number of oil companies and welded a portion of Tennessee Gas Transmission's twenty-four-inch pipeline during World War II.

The FPC, on May 31, 1946, granted the application. To finance the cost of the project, El Paso issued new first mortgage bonds and preferred stock, contracted a bank loan and offered 100,057 additional shares of common stock for holders of the Company's common stock. Financing of the project was completed on July 2. On July 23, El Paso placed $42,212,000 cash in the bank as a construction fund for completion of the huge project.

Construction of the pipeline kicked off in February, 1947, involving the laying of more than 700 miles of pipe and the building of compressor stations necessary to keep the gas moving to California.

With a deadline of less than a year to begin its first deliveries to the California distribution companies, El Paso's work force literally turned upside down. During 1946, names on the payroll increased fifty percent.

R. W. Harris, then division superintendent at Jal, recalled that he would work all day at Jal and then drive into El Paso at night. "Perkins and I worked on what equipment we needed," said Harris. "We'd decide what the Purchasing Department should order the next day, and then I would drive back to Jal that night. I did that for a number of months until we got all this equipment. Mr. Kayser gave Perk the authority to go ahead and buy electric welding machines and welding rods, and we gave the Big Three Welding Company in Fort Worth the largest single order they had ever gotten.

"We could feel the excitement in the air, but nobody really realized then what a big job it was and what an undertaking we took on. But we felt that with the experience we already had, we could take boys returning from service who really wanted to work, and we could get this job done on schedule. We also hired some old pipeliners who knew their stuff.

"We actually revolutionized the pipeline industry. We developed a way to smooth-bend pipe that is still used today. We modified the ditching machines. We were the first ones to use the big pipeline ripper. We did things that had never been done before, and we set national and international records in laying pipe. Before our California project, anybody that could lay 4,000 feet of big pipe in a day—as far as getting it in the ditch and coated—was doing real good. Well, we laid as much as 15,000 or 16,000 feet a day. We did that through necessity. We had to do it."

Hank Schulze remembers how it was when El Paso Natural was tooling up for the main twenty-six-inch line to California. Says Schulze, "We had every pipeline phony and every has-been comin' and goin' by the dozens. Some of 'em were absolute sheer geniuses if you'd listen to 'em and they'd go to work and maybe they'd last a week or maybe they wouldn't last that long.

"But we were pickin' up everybody that came by that had any pipeline experience. This was gonna' be one of the biggest jobs that had ever been pulled off in this country, so we did need a lot of help. They got me a kind of advisor, a guy by the name of Williams, and he did have some good ideas. But he had so many ideas that the bad ones overrode the good ones. He lasted about two or three months, maybe four months. He let a section of pipe get loose and it went down the side of a mountain, and when it got to the bottom, it made an accordion out of about 150 or 200 feet of pipe. So he went off down the road kickin' a tin can and talking to himself."

The first use of large pipe was on a test section laid through Anthony Pass, north of El Paso. The Company chose the site because of the ruggedness of the terrain, with inclines reaching angles of forty-five degrees and places where the ditch had to be blasted through solid rock. Experience gained from bending and doping pipe through this stretch proved to be profitable, since it served the pipeliners well in familiarizing them with the type and ability of the heavy equipment that would be necessary in a project the size of the California line.

John Eichelmann, who was chief engineer at the time, recalls that the size and weight of the big pipe was truly a new experience, but trying to weld the sections together bordered on disaster. "Because of the shortage of steel pipe right after World War II," says Eichelmann, "we were forced to use some very high-carbon, high manganese pipe for this job, and we just couldn't weld it by normal methods. In desperation we decided to try our old 'pressure welding' machine."

The "pressure welding" method came about during the early Forties, developed by El Paso and Linde Air Products. A. L. Forbes initiated the project which eventually became known as the Pressure Welding Company, a subsidiary of El Paso Natural. It was a unique method of joining pipe — without the use of welding rods. Very simply, it used a circular burner capable of creating extremely high temperatures. The ends of two lengths of pipe were heated, and when they reached fusion temperature and pressure was applied, the pipe was literally fused together.

El Paso first used pressure welding on a six-inch pipeline from Phoenix to Tempe, Arizona, in 1940. Mildly successful for awhile, the system had its drawbacks. According to Eichelmann, when El Paso pressure-welded three miles of large-diameter pipe on the test section through the roughest part of Anthony Pass in the summer of 1946, it just wouldn't work on the big pipe. But it did a fine job on anything ten inches or less in diameter. "After pressure welding on the test section," says Eichelmann, "we filled the new pipe with gas at high pressure, and breaks started occurring by the dozens. Actually, much of the trouble developed because we were using seamless pipe, which didn't have a uniform wall thickness. We knew it wasn't going to work so we dug it all up."

In the meantime, Perkins was getting desperate for a solution. And the solution came from Hugh Steen and the No. 3 machine shop in El Paso. After a number of false starts, Steen and two top welders, with the assistance of Lincoln Electric Company, developed a method of using "stringer beads" and a "hot pass" to hold the pipe together before making final welds. This provided the extreme heat needed to prevent quick cooling and although it took talented welders to handle the method, it did the job beautifully—even on the high carbon steel pipe.

What did it take to build a pipeline from Texas to California? It took the combined skills of several thousand people. It combined the creative talents of engineers and mathematicians working without the mechanical speed of today's electronic computers. It took mountains of paperwork and executive planning. It took lawyers and right of way agents and secretaries and many more. But in the final analysis, it took the pipeliners and construction crews living in close contact with bone chilling cold and blowtorch sunshine. They fought the wind, the sand, and blowing dust. They were on speaking terms with the worst that Mother Nature could provide.

It was a long way from the Texas Panhandle to the Pacific. And although much of the distance followed pioneer trails, the pipeline blazed a new one across the West. On a map it looked easy; simply a line from here to there. But from the air its immensity became more apparent. Hundreds of obstacles blocked the way—sand, deserts, rivers, mountains of solid rock. Before construction could begin, skilled engineers spent thousands of hours in the drafting room.

According to John Eichelmann, the Company went into the California project with a tiny engineering staff. "But you don't just handle a job like that without extra hands. We had Lambert Moore and Hector Valencia and a couple of others in the drafting room, and that was it.

"And then we started expanding. We hired an awful lot of hard working boys for the engineering department, about fifty of them. Along with engineers out in the field doing survey work, we had a total of about 125. From there on we multiplied to a great engineering group. We were quite proud of what we accomplished. I know of a lot of engineering firms that were amazed at the amount of work we did. It just shows that if you have the will and the enthusiasm you can't keep from doing a good job. There wasn't anybody who held back because it was five o'clock.

PIPELINER Hank Schulze
(left) was one of hundreds of
El Paso hands who helped
revolutionize the pipeline industry.

THE FIRST California line
construction began in February,
1947 and involved the laying
of more than 700 miles of large-
diameter pipe.

We'd go over and eat at seven, come back at eight and work 'til midnight and then start all over in the morning. And it wasn't that the Company was demanding. It was the fact that everybody had enough interest. I think it's something this Company can be proud of, to keep people with that kind of enthusiasm."

After engineers planned the project on paper, preliminary work remained in the field, like rights of way secured from hundreds of individuals and firms. Engineers determined the exact route to achieve a balance between shortest distances and lowest construction costs. Much of the survey was done from the air first, then laboriously covered on the ground, making sure the chosen route was practical.

Much of the pipe used was fabricated on the West Coast, and in spite of disastrous shortages, the steel companies and fabricators in the East and West caught the spirit of this tremendous undertaking. By almost superhuman efforts they made deliveries so the job could be finished on time. Manufacturers couldn't supply side-boom tractors capable of lifting the heavy pipe, so the Company bought sixteen Cletrac farm tractors and modified them with wider tracks and heavy weighted devices to hold them upright.

Pipe and fittings were transported by rail to stockpiles at strategic points along the route. El Paso Natural invested more than a million and a half dollars in new equipment before the first piece of pipe could be laid. Along the right of way, tractors, bulldozers, welding equipment, trenching machines and a thousand and one other items arrived from all over the country.

Four "spreads" or pipeline crews, each composed of about 250 men, kicked off simultaneously clearing the right of way. Spread No. 1, under the supervision of Jack Smith, began constructing gathering lines in West Texas and New Mexico and moved westward across the Pecos to the Guadalupe Mountains. Spread No. 2, under Gaston George, started through the rim rock section of the Guadalupes and moved west to near Deming, New Mexico. Ed Alsup was superintendent of Spread No. 3 which went from Deming to the Tucson, Arizona area. R. W. Harris later headed up this spread. The fourth, a contract crew from Midwestern Constructors of Tulsa, Oklahoma started laying pipe in the Casa Grande Valley moving westward to the Colorado River. E. W. Woody headed up a "short crew" to handle certain rough spots in Texas Canyon, Apache Pass and the Dome Mountains in Arizona. C. L. Perkins was in charge of all the pipeline crews.

After trucks hauled the big pipe to the right of way, pipe sections were strung end to end in preparation for the next step in construction. In remote areas, tent camps were set up for the pipeliners with basic essentials to provide a measure of comfort for the men.

In the meantime, treating plants and compressor stations had to be constructed from the ground up so that gas could be flowing through the line by the fall of 1947.

It was a tight schedule calling for real teamwork and a friendly competitive spirit among the spreads. It was rough going. Ten or more hours a day. Seven days a week. Week in, week out; month in, month out; in good weather and bad.

Sometimes with good luck, they made a couple of miles in a single day. Other times when nature and balky machinery conspired to whip the men down they would go only a thousand feet. In the soft earth of irrigated land, the big ditching machines did a speedy job of making a trench four feet wide and eight and a half feet deep. Most places required a ditch only three feet wide and six and a half feet deep.

After the trench was dug, the line-up crews took over, and they provided one of the best examples of teamwork in pipe-laying. A clamp inside a forty-foot section of pipe held a new section in exact position until a tack weld was completed. The clamp was released and moved ahead to align another section. Then came the final welders who completed the weld on each section of pipe. This is a delicate job requiring good judgement, skilled craftsmanship and no compromise with perfection.

Construction on this first California line proceeded much the same as it is being done today: after the line is welded together, the outside is brushed clean and primed with a corrosion-inhibiting material. Then the line is wrapped with various types of fiberglass, plastics, coal tar and papers. This job, too, is carefully supervised. An electronic instrument is then run over the entire line to check it for any possible flaws in the coating and wrapping. The device can find flaws too small for the human eye.

The next step is lowering in. Big side-boom tractors gently placed the wrapped pipe into the trench, usually in the early morning before the metal expands from the heat and the coating becomes soft. Backfill crews take over and use bulldozers and maintainers to replace the soil removed in digging the trench. In rock ditch areas, the pipeline is padded with soft earth or sand before the final backfill is pushed over to cover and protect its coating from rocks or other obstacles. The right of way is then leveled, fences rebuilt and the land returned — as closely as possible — to its original state.

On the California line, El Paso's pipe-liners ran headlong into some of the most formidable obstacles they ever encountered. Like the Rim Rock escarpment below Guadalupe Peak in far West Texas. The pipe had to be laid 1,600 feet down the base of a rocky cliff at an angle of forty-five degrees. Using winches anchored on top of the escarpment, they lowered the tractors down the cliff with cables to scrape out a semblance of a road, to dig the trench and to inch the long lengths of pipe into position.

At the foot of the Guadalupes are the ruins of the Butterfield State Station at Pine Springs, a reminder of other obstacles faced by earlier pioneers, not all of whom reached their destination. At the bottom of the peak is the rough grave of one who fell by the wayside, killed February, 1855, by Apaches.

A few miles to the west was another obstacle of an entirely different sort; the famous Salt Flat where the precious salts scraped from these natural beds caused the bloody and bitter Salt War of 1877. Water, seeping from the earth, carries with it the salt that gives these flats their name. Just below the surface is a layer of black, gummy, sticky mud—no footing for even a versatile tractor. So the equipment worked from a base of heavily timbered pads, twenty-four feet long and twelve feet wide. Once the trench was dug, the pipeline spread prayed for dry weather so that rain wouldn't undo their labor. But here in the shimmering August heat of 1947 the crews worked with little concern for their unfriendly surroundings.

Not all of the obstacles along the pipeline were rocks and mountains and miles of salt flats. There were rivers, too. Some of them, placid streams in times of normal flow, turned into raging torrents during flash floods; the Pecos, the Rio Grande, the San Pedro, the Gila and at last the mighty Colorado.

Bridges, some of them 2,000 feet long, had to be erected over the rivers and adjoining flood areas. These suspension bridges are of peculiar construction. Wide arms on the towers support cables that brace the pipe against side swaying. The pipe was raised to work platforms and welded. A tractor pulling a cable moved the pipe inch by inch toward midstream. On the far end, a pipeliner wearing a safety belt attached to one of the supporting cables guided the pipe into cradles as it was pushed forward.

PIPELINE SPREADS often encountered 20 or 30 rattlesnakes in a morning's work on the way to California.

BUILDING A NATURAL GAS PIPELINE: THE MAJOR PHASES

1. After the right of way has been cleared, ditching crews follow, cutting a trench deep enough to provide approximately two and a half feet of cover over the pipeline.

4. Sections of the pipeline are cleaned, covered with primer, "doped" with a protective coating, then wrapped snugly in a jacket of felt to prevent corrosion while in the ground.

5. An electronic sensing device checks the coating for imperfections, called "holidays." Small holes that may appear in the pipeline coating are marked by the crew and enamel is applied to the pipeline by brush.

2. The pipe is then bent to the proper shape to fit the terrain, and the next step in the construction is the lining up of individual lengths of pipe alongside the ditch so that each length can be connected together by means of a "tack weld."

3. Welders carefully join the joints of pipe together in welds stronger than the pipe itself.

6. Lowering in. Big sideboom tractors gently place the wrapped pipeline into the trench, usually in the early morning before the metal expands from heat.

7. After the backfill crew replaces soil and buries the pipeline, the right of way is leveled and returned—as closely as possible— to its original state. Then every mile of the line is periodically patrolled by truck, on foot, or from the air.

AT THE ESCARPMENT below Guadalupe Peak, big pipe was
laid 1,600 precarious feet down the rocky slope.

PIPELINERS worked at angles
of forty-five degrees in some
spots coming off the rimrock.

THE ROCKY FACE of
El Capitan looms above
construction equipment on the
California line. At 8,751 feet,
Guadalupe Peak (center right)
is the highest point in Texas.

In Arizona, Spread No. 3, was fighting tough spots of hard, solid rock in Texas Canyon and the Dome Rock Mountains. Almost every foot of the trench had to be blasted out of granite, shale or limestone. Rocky soils, steep grades, and many sharp curves provided obstacles that couldn't be penetrated by conventional ditchers, contributing to the difficulty of the task. Men with jackhammers had to drill holes, although in some places automatic drilling machines could be used. Pipeliners placed dynamite, laid fuses and blasted out a hundred feet or so of trench at one time. After the last of a series of explosions rumbled nearby, the heavy tractors scraped out the rock and rubble. No mile-a-day progress here. Arizona's Apache Pass didn't give up without a fight either. Hank Schulze says, "We got up into Apache Pass and it was rougher than the back end of a shootin' gallery up there, I'll guarantee you that. It was rough, rough, rough.

"The only good thing about that whole job was the lunch we had every day. We had one old man that would stop in town every night and buy up a bunch of groceries, and he'd go up ahead and build a fire and on the coals he'd set up a big dutch oven full of stew. Well, we'd make to that spot around noon, and that stew was somethin' else.

"We used to run into a few rattlesnakes, but the only time they'd give you any trouble was if you had a little cave-in or something in a ditch. You wanted to be sure you looked real good in the ditch, 'cause when we went through the top soil maybe three or four feet down we'd hit a sandy layer. Well, the pack rats, the kangaroo rats and mice runnin' up and down the ditch would get this sand to cave in right under the top soil and leave a little shelf down there. These snakes would get up under that shelf to get out of the sunshine, and if you popped down in there you could be standin' within two or three inches of them. One mornin' we hit a litter of little rattlesnakes and we must have killed sixty or seventy of 'em in a twenty-foot section right there. They were just everywhere. We just took a bucket of hot dope and went down the middle of the ditch, just doped 'em where they were down in the ditch. A lot of times it wasn't unusual to run into fifteen or twenty snakes in a mornin's operation. A lot of this operation went on before daylight and we had no idea what was down there. We had lights on the tractors so the operators could keep from hittin' the sides of the ditch with a pipe when they were lowerin' it in, but you couldn't see down in the ditch at all. That'll drive you to drink, I'll guarantee ya'."

While pipeliners continued their race to complete their part of the job, mainline compressor stations were being built; the first ones at El Paso and Tucson to take care of the initial load. Then others followed; Guadalupe, Deming, Willcox and Gila. Crews built homes at each station to house employees and their families. Although some of the "camps" were miles from the nearest town, they became modern, pleasant little communities.

GUADALUPE Compressor
Station moves Permian
Basin gas westward to
California and points
in between.

SPOTLESS INTERIOR
of Guadalupe Compressor
Station is typical of El Paso
Natural's installations.

One of the first jobs of the new compressor stations was to blow out sections of the lines, many miles long, with gas under pressure cleaning out the dirt accumulated during construction and purging it of air. Then welders made the final connections. In the gas fields of the Permian Basin, the building of gathering lines, treating plants and compressor stations progressed simultaneously.

C. C. Selley was in charge of the construction of all plants and compressor stations on the California project, and a number of construction hands like Norris Stogner, Baldy Henderson, H. C. Doan, Bill Crutcher and others went on in later years to become vital links in El Paso's chain of command. Walt Miller was superintendent of all compressor stations in the system.

Doan remembers the building of Willcox Station when the pipeline crews were nearing Willcox. "They were going through all these ranches," says Doan, who later became the superintendent of the Southern Division, "and they were supposed to close all the ranch gates to keep cattle from getting out. One night this spread was in a hurry to shut down the job and they left the gate open by mistake. The next morning the rancher was out there looking for the man that was running the job. It was R. W. Harris. Well, he found R. W. and said, 'You all let one of my bulls get out and he's lost. I can't find him.' So Harris said, 'Aw hell, don't worry about that. We'll pay you for him.' And the rancher said, 'Well, if you've got $45,000 you can

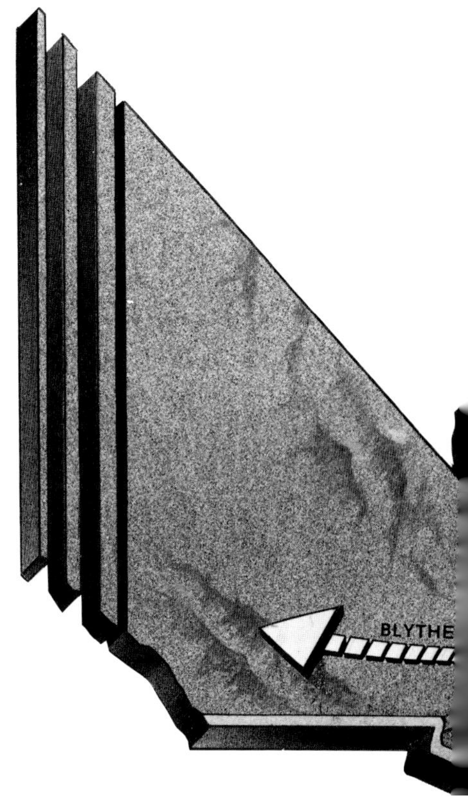

BY 1948, the pipeline and plants to serve southern California customers had been built, extending from Dumas in the Texas Panhandle to the Arizona-California border on the Colorado River. Meanwhile lateral lines and extensions were built on the original line to serve communities in southern New Mexico and Arizona.

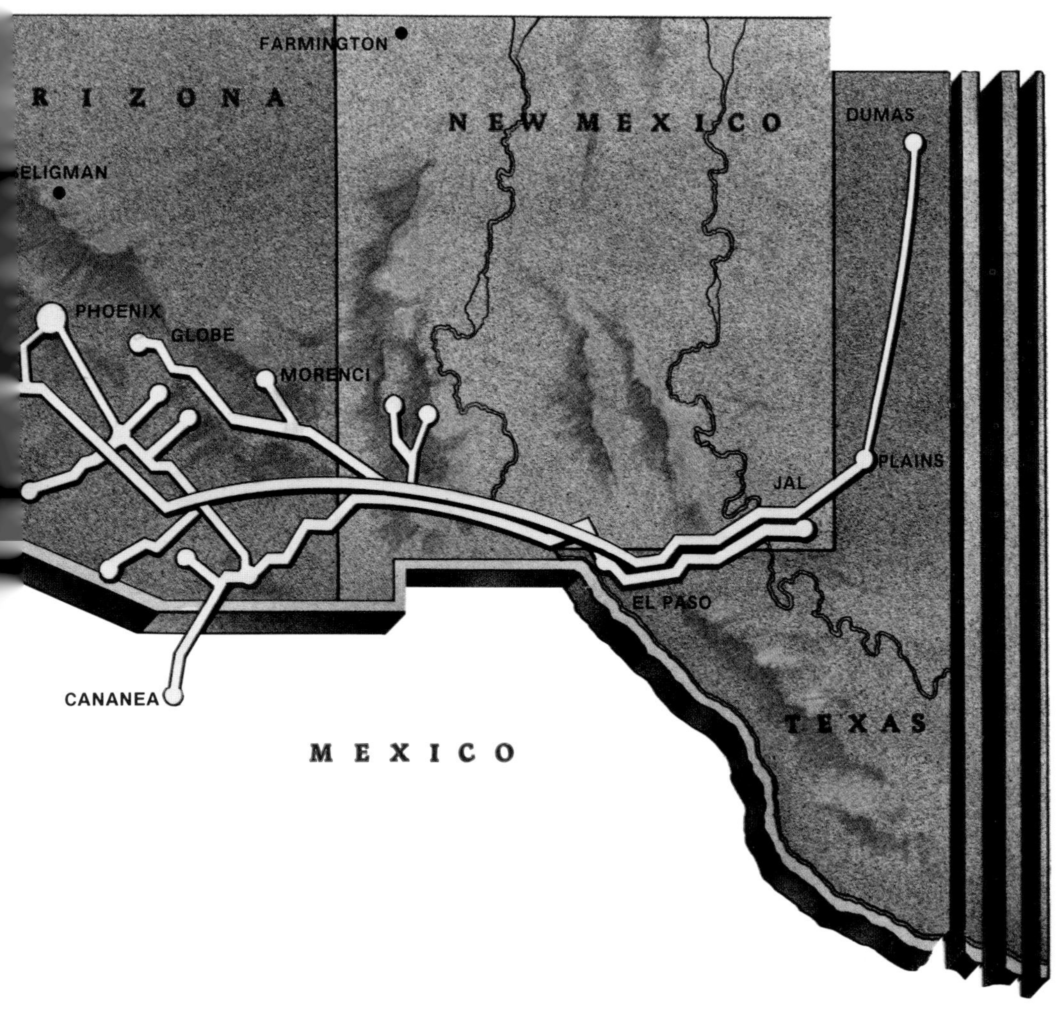
ARIZONA
NEW MEXICO
TEXAS
MEXICO
FARMINGTON
SELIGMAN
PHOENIX
GLOBE
MORENCI
CANANEA
EL PASO
JAL
PLAINS
DUMAS

pay me for him!' So Harris shut down the whole pipeline spread and they went out and found that bull. It turned out to be a prize bull and they spent more than half the day looking for him."

Through the early fall of 1947 construction crews sprinted toward the finish line, the Colorado River, for the hookup with Southern California Gas Company and Southern Counties Gas Company. The California companies completed their part of the project from Blythe, on the Colorado River, into the outskirts of Los Angeles. Construction men applied finishing touches to compressor stations and treating plants, checked every detail and made final tests. Then the natural gas from Permian wells and gasoline plants entered the new line. Valves were opened to divert more gas into the treating plants. Gas that had been wasted to the air for more than two decades became valuable fuel as hundreds of giant flares flickered out in West Texas and eastern New Mexico.

Right on schedule, on November 13, 1947, at impressive ceremonies held at Santa Fe Springs, California, Goodwin Knight, Lieutenant Governor of California, turned a valve that started the gas racing into California after its eighty-hour journey from West Texas.

Said Paul Kayser at the inauguration of the gas flow to California: "I wish to acknowledge the debt of deepest gratitude to the whole generation that has taken on this job, under the most adverse conditions — and has performed this job in record time, and yet maintained an integrity of performance that assures the efficiency of the operation of the line throughout its life. This was no mean task, and my hat is off to everyone who has performed his service on the project."

Hugh Steen puts it this way: "On this pipeline we learned a lot about coating pipe, soil erosion, corrosion and electrolysis. We had our part in the development of all of these things, and I say that very modestly because we certainly didn't do it all. But we certainly did a lot. We had some old boys that were great and a lot of 'em are still here. You take Ed Alsup, Hank Schulze, Gaston George—guys like that were responsible for the thing getting through. But the driving force was Perkins. And of course, R. W. Harris was the one who ran it, the pipeline spreads. I can't say enough about any of these people."

The completion of the California pipeline soon proved to be only the beginning of the great expansion of El Paso Natural Gas. The Company's newly employed Home Office personnel overflowed the Bassett Tower and the engineering offices in the El Paso Electric Company Building. New employees moved into the old El Paso National and Southwest National Bank Buildings and others. In 1947, the Company organized a transportation department with a total of 305 vehicles.

Even before completion of the twenty-six-inch line, the Company applied to the Federal Power Commission for permission to build 423 miles of loop line to California. The line would supply an additional 100 million cubic feet of gas a day to Southern California.

By 1948, El Paso's rapid growth caused

PAUL KAYSER, then president of El Paso Natural, speaks during ceremony at Santa Fe Springs inaugurating gas service to southern California. The coming of Texas gas to California was right on schedule; November 13, 1947.

END OF THE LINE. Natural gas from Texas crosses the Colorado River
near Blythe, California. This picture was taken from the California side looking
east to Ehrenberg, Arizona.

some near seam splitting. New demands for gas all over the Southwest created the need for more expansion in mainline and field stations. Pipeliners laid more than 600 miles of pipe, including a seventy-four-mile line to Yuma from the twenty-six-inch line at Quartzite, Arizona. They laid this line almost due south to Yuma, to serve, among other industries, an alfalfa dehydration plant there.

The blast furnace heat on this job stuck in the mind of R. W. Harris, who by that time was pipeline superintendent over the whole system. "Every day the temperatures hit 118 or 120," recalled Harris. "And I remember that we didn't take many of our pipeline men out there because we thought it would be better to hire the local people as they would be used to the heat. Well, we hired the locals but they began to drop from heat exhaustion, but our boys could withstand it because they had been working out in the desert for years. Come to find out, the local men had always done their labor and plowing at night. They just didn't work out in the heat of the day,

148

which goes to show you that a bunch of pipeliners didn't have very good sense.

"We learned pretty quick, though, that you couldn't pick up a wrench or any kind of a tool without burning your hands. So everybody wore gloves and we learned to dress like Arabs. We learned to put *on* clothes, not take them off to keep cool. The welders wore normal clothing including khaki shirts, and on top of that they put on overalls and jumpers. Big, long leather gloves and their helmets helped shield them from the heat."

The expansions in 1948 called for more housing, and 200 houses were built at various stations and plants. This meant more Home Office personnel to keep track of everything. More than 400 employees hired on during the year, making the total number in construction and operations nearly 2,500 — tripling the employee roster in three short years.

With all the new faces around, the Company formed a Personnel Department under the direction of Ralph Allen. Burney Warren, who was later to become Executive Vice President and a Director of The El Paso Company, received a promotion to chief clerk in the Accounting Department.

All the growth was caused by the system's steadily increasing delivery capacity. In the early part of 1948, capacity stood at 345 million cubic feet of gas a day. Construction during the year raised that to 511 million cubic feet.

By 1949, the demand for natural gas continued to increase at an unprecedented pace, and it seemed unlikely that the late-burning lights in El Paso's Engineering Department would be turned off in the near future.

The Company acquired a site in downtown El Paso for a badly needed Home Office Building which would bring together the widely scattered forces then being housed in offices all over the downtown area. Walking from some offices to others took on the proportions of a major field trip.

The Company's air arm, under chief pilot A. M. "Swede" Johnsen, flew over a million and a half passenger miles a year, more than some small airlines. Oldtimers in the Company, who remembered those lonely and tedious drives across the desert from station to station along the line, now stepped into a Company plane and often within minutes bounced down a landing strip bulldozed beside a new plant or compressor station. 1949 also saw the completion of four new plants in what was then called the Jal Division: Dimmitt, Goldsmith, Keystone Mainline and Wasson.

As the 1940s drew to a close, El Paso's employees could look back on a decade filled with major accomplishments. From a work force in 1940 of 413, the operations employees now numbered 1,281. (This did not include employees on the construction payroll.) The character of the Company had changed from that of a small regional pipeline company to a major corporation. Importantly, from the standpoint of conservation, thousands of oil well flares disappeared forever in the Permian Basin.

And waiting backstage was a spot that would soon become a household word in the El Paso family—the San Juan Basin.

8

THE SAN JUAN BASIN

Next stop, San Francisco

THROUGH the timberland of northern Arizona, Navajo Indians played a major role in construction of the San Juan pipeline.

THE LITTLE TOWN of Farmington dozed quietly under warm New Mexico skies. Main Street loungers conversed mostly about the weather, the fruit crop and an occasional unfamiliar red pickup truck with El Paso Natural Gas Company markings driving down the street. The date: 1950.

Six years later the face of Farmington was almost unrecognizable. The "unfamiliar" red trucks, vanguards of hundreds of El Paso Natural vehicles, rumbled in to become part of the Farmington scene, as more than a thousand El Paso employees helped swell the city's population to more than 16,000 by 1956.

The cause of this incredible growth in Farmington was its fortunate location at the hub of the gas-rich San Juan Basin of northwestern New Mexico. As early as 1911, oil had been discovered here by water well drillers, but demand did not justify pumping it out of the ground.

It took the post-war boom in California to change all that. And as one old-time resident puts it, "With the coming of El Paso Natural in 1950, the lid blew off."

Actually, plans for taking gas from the San Juan Basin, as well as additional supplies from the Permian, began taking shape even as the ink was drying on the contracts and the dust was still settling on El Paso's first pipeline to California from the Permian in 1947. Paul Kayser announced on August 26, 1947, a new construction project designed to make El Paso one of the largest pipeline companies in the world. The Company would build 423 miles of

151

loop line to parallel the original California pipeline and supply an additional 100 million cubic feet of gas a day to southern California, carrying residue gas that was still going to waste in the Permian Basin.

Pacific Gas and Electric Company signed contracts for delivery of some 300 million cubic feet of gas per day to San Francisco and northern California. The new San Francisco line would eventually tap the Permian Basin fields as well as the San Juan Basin. A 450-mile, twenty-four-inch line from the San Juan would stretch through the Navajo Indian Reservation to Topock, Arizona—just across the Colorado River from Needles, California. Also, the new southern and northern lines would be connected by a thirty-inch crossover line running slightly over 100 miles from near Ehrenberg north to Topock, carrying Permian Basin gas.

El Paso's civil engineers surveyed the new San Juan line. The shortest practical route was determined first—this time by air. Then rights of way across the land were secured from hundreds of property owners as well as the Federal Government. About half the route crossed the Navajo Indian Reservation, and meetings began with the Tribal Council at Window Rock, Arizona. From all over the Reservation (largest in the nation), Navajo leaders came to their council chamber for important sessions with El Paso and government representatives.

As in many of El Paso's pioneering efforts, one man stood out in making the San Juan project a reality: C. L. Perkins. The personable, tough taskmaster knew his

C. L. Perkins

business. He also knew how to get along with people — from laborers to business executives.

"At that time," says Perkins, "P. V. Fuller was in charge of our Right of Way Department, and he made the first contacts with the Bureau of Indian Affairs and various government agencies. But this job was something new for us; the forest lands we had to cross and our dealings with the Navajo Indians. I decided the best thing to do was to go along with Fuller and Ben Howell, who was an outside lawyer who was doing work for the Company. (Howell later became a Vice President and a member of the Board of Directors.)

"We met with the advisory committee of the Navajo Tribal Council, it being easier to meet with its nine members than the entire seventy-three people. The advisory committee could act for the Council and their actions were usually ratified at the next meeting.

"We went up to Window Rock and explained that we wanted to acquire 151 miles of right of way across the Reservation, and we quickly found out that it was necessary for all of our statements and questions to be translated into the Navajo language. They spoke in Navajo and it was translated into English for our benefit. We were very fortunate to have an excellent translator, Paul Jones, who later became Tribal Chairman. Sam Ahkeah was Tribal Chairman at that time. In two days of discussions, they asked some very searching questions that showed the Navajos wanted to improve conditions on their Reservation. We felt like everything we did would be of benefit for the Tribe.

"Naturally, we had a few problems training the Navajos to become pipeliners, mostly because of the language barrier. The Navajo language is very descriptive, so they learned by watching or being shown how to do things. A lot of them had already learned to drive bulldozers in the Army during World War II. We also used half tracks with welding machines mounted on them, and the Indians quickly learned how to maneuver them with skill. Thanks to people like Jesse Williams, who worked for the Company in Farmington and knew many Navajos, he accomplished things that few others could. He had been County Judge of Hudspeth County, Texas and also knew about everybody in the Farmington area. We used bilingual Indians as foremen and personnel people and the whole project moved along smoothly. Ben Howell and Jesse Williams really did a yeoman's job up there on the Reservation. It was a credit to all our people that such pleasant relations resulted."

On the international scene, the world had come through its second great war in a generation to find that peace, of the old fashioned kind, didn't exist, as the cold war replaced the hot war. Behind closed doors and iron curtains the world was playing with devices which could write finis to the history of mankind in one blinding flash.

The Bikini nuclear explosions, the Berlin airlift, the million dollar robbery of Brink's in Boston, all briefly headlined the front pages of newspapers and a new

DRILLING FOR GAS in the San Juan Basin.

medium called television. Then came a dateline which would be around for quite awhile: Korea.

El Paso Natural employees, many of them "retreads," as they called themselves, were recalled into the service to help stem a North Korean communist army in its invasion of South Korea. What began for U.S. troops as a police action soon became a full-scale war, and for the second time in a decade *The Pipeliner* was running letters from servicemen, this time serving in the Korean conflict.

El Paso's San Juan Project received a fast green light on high priority steel pipe since West Coast industry would carry a heavy load for new armament to fight the war. It would be far better to have the San Juan Project in operation rather than under construction should the war escalate into another world-wide conflagration.

Over rugged mountains, across the Mojave Desert and the fertile San Joaquin Valley PG&E's line was rushed to help supply the ever-increasing demand for more gas in northern California. The discovery of great new reserves in the Permian and San Juan basins coincided with the expanding demands of the Southwest and San Francisco areas. Now both were to have a continuing supply of natural gas. It was possible to build and finance the San Juan line because of northern California's vast market and at the same time make possible deliveries for the first time to communities along the route. Gallup, New Mexico, home of the Intertribal Indian Ceremonial; Holbrook, Arizona, with its painted desert and petrified forest; Winslow, railroad and trading center, and Flagstaff, with its lumber mills at the foot of San Francisco Peaks, were to get natural gas, as was Prescott, the largest city in northern Arizona. The pipeline would also serve Williams, the gateway to the Grand Canyon, as well as Ashfork and Seligman, and Kingman, mining town and tourist turnoff to Hoover Dam and Las Vegas.

R. W. Harris recalled a brief conversation with C. L. Perkins, who was then Vice President and general superintendent of the Company. "Perk said, 'Alright, you go up there and build this pipeline and take the pipeliners, all our good loyal Company men, and fit them into an operational program.' So we went up there and started this pipeline on the day after Labor Day in 1949.

"We did some pioneering in labor relations up there, too," said Harris. "Hundreds of the pipeliners on this job were Navajos and they picked up the tricks of the trade in short order. We gave them equal jobs, equal opportunities for themselves and their children—which they had never had before. Perkins established that tradition, and later other companies and other projects followed right along this same line."

In those days, most people referred to Harris by his nickname—"Lazy." Perkins says that Harris picked up the name in high school and it stuck for many years. "We were all in high school in Clyde," says Perkins, "and we had a football coach named Pat Murphy (who later coached at Austin High School in El Paso). During our workouts, he'd have us run around the field doing sprints. We didn't have any stands or anything. Well, ol' R.W. couldn't run fast enough to get hot, and Pat hollered at him and said, 'What's the matter with you lazy?' And, of course, ol' R.W. was runnin' just as hard as he could, but he got stuck with that nickname. I think that's one thing that made him such a great man. His name was Raymond Weldon, and his wife, Alice, would call him Weldon. Mr. Kayser wouldn't call him "Lazy" and didn't like callin' him Weldon, so he got to callin' him R.W. But to all the pipeliners and everybody he was known as Lazy Harris. As the years went by, though, more and more called him R.W. He was anything but lazy."

Meanwhile, crews raced against time to complete certain parts of the San Juan pipeline before winter closed in, as San Juan winters can build into blizzards overnight that hold temperatures near zero for days at a time. The terrain varied from prairie to mountain, from sand to solid rock, from desert to forest land. Over 160 miles of the line had to be blasted through solid rock. The job required more than two million pounds of dynamite.

To comply with the Antiquities Act and to preserve any relics of ancient civilization along the line, the National Park Service people accompanied the right of way and trenching crews. El Paso cooperated with the government by putting these archeology crews on its payroll. Part of the Company's team on the project worked under Dr. Jesse L. Nusbaum, senior archeologist of the National Park Service. These scientists recovered and examined the valuable artifacts unearthed along the line. As the San Juan line entered the Navajo Reservation, the archeological crews which were working in front of the construction spread hit an area rich in ruins and relics of great value. Lots of buried pit houses and other prehistoric Indian relics were unearthed along the right of way across the Reservation.

Outside experts recommended the use of professional lumberjacks to clear the right of way through forest land. But El Paso decided that its construction department needed no outside help. The "desert rats" quickly familiarized themselves with handsaws, power saws and chain saws to get the job done on time. Navajos wearing their traditional long hair and turquoise jewelry worked alongside the new lumberjacks.

In its original concept, the San Juan pipeline and facilities were to be used only during the winter months to supplement gas coming westward from the Permian Basin. However, it never really worked that way, since the demand for gas in northern California became so critical the San Juan Basin poured forth its bounty of energy as fast as the new customers could accept it.

R.W. Harris supervised the building of the Company's San Juan Division, and in 1950, he asked for some top hands to help run the plants, pipeline, compressor stations and related departments. He naturally looked in the direction of the Permian Basin, and one of the first names he jotted

IN THE FORESTS of northern Arizona, this Navajo crew aligns two lengths of 34-inch pipe during construction of the San Juan pipeline.

DITCHING MACHINE near Flagstaff, Arizona makes progress along the right of way from San Juan to San Francisco.

down was Frank Galle, plant superintendent at Jal No. 1.

Galle remembers it this way: "R.W. said, 'Frank, we're gonna' build a pretty nice plant up there; gasoline plant, compressor station, treating plant. But I want to tell you one thing, though, before you consider it, we're only gonna' operate about six months out of the year. And we'll be shut down during the summertime. During the summer you're gonna' have to take your operating crew and go down in the Southern Division and overhaul engines. Then when winter comes, we'll run the San Juan River Plant again.' " (What Harris was referring to was the original certification from the FPC. The Commission felt that San Juan's gas supply was not adequate to operate at full capacity for more than six months a year.)

"I came on up" says Galle, "and helped build the plant. We put it in full operation sometime in September of 1951.

"That shutting down for six months in the summer never came about. We never even slowed down; we just got stronger. They kept drilling wells, and found that there was a lot more gas in the San Juan Basin than most people had originally thought. We started out up here operating the San Juan Plant on gas from the Barker Dome Field, which was a sour gas field. Those were large wells. There were only a few of them — about seven or eight — but they would produce a lot of gas.

"The Barker Dome Field, of course, has pretty well played out now, although it will be producing gas for many years. Once we were taking something like 140 million cubic feet a day out of those few wells. But for the last several years we've only been taking out about six million a day. But it's holding up.

"Nowadays I believe we're tied-in to something like 6,400 wells all over the basin. So you can see what a large field it turned out to be. I only wish the field would do today what it would just a few years ago."

The San Juan River Plant project soon surfaced as a giant migraine headache to everybody involved in its construction. The massive stills, contactors and other pieces of equipment were too heavy to be hauled on the highway from the railhead at Gallup to Farmington.

There are 137 miles and forty-eight bridges between the two communities, and unless you are in the highway or bridge construction business the prospect of moving a 180,000-pound load over the desert is not particularly inviting. Loads like this will quickly convert bridges into heaps of twisted metal and splintered wood, while the pavement becomes eligible for participation in the next dust storm. So it fell to El Paso's Building Construction Department to solve the problem of moving these crushing loads.

The only way the job could be accomplished was to move the big pieces of equipment, some of them sixty feet long and eleven feet in diameter, along the pipeline road. Sandy, dirty, isolated, the pipeline road differed from the surrounding desert country chiefly in that some sparse vegetation had been cleared away. The road itself was not paved, graveled or otherwise improved.

Building Construction crews put together a huge trailer equipped with tracks cannibalized from worn-out tractors used in the construction projects of bygone years. The giant steel towers were loaded onto the trailer and fastened securely. Three bulldozers (and sometimes four) moved the trailer along the pipeline road.

All supplies and material necessary to keep the dozers and their load moving had to be carried along. The procession included a half-track carrying drums of fuel, grease and a welding machine; a grease truck with machinery to lubricate and refuel the equipment, a heavy winch truck for lifting or moving machinery; two automobiles to pick up personnel or to get parts.

At best, a track-laying tractor is a heavy, slow-moving, jarring monster — probably the world's most uncomfortable method of transportation. Hitched to their cumbersome load, the cats moving the trailer made not over a mile-and-a-half an hour. The caravan operated on a twenty-four hour a day basis, changing crews at six in the morning and six at night. A single trip required two or three days.

Every thirty minutes the men greased the tracks and bogie wheels (bogie wheels are the small wheels rolling within the tracks) of the tractors. Occasionally, a bogie wheel heated up and caught fire.

The large trailer broke down repeatedly. When a bogie wheel pin sheared or some metal part cracked, welders quickly repaired the equipment. Fuel pumps went out, bearings overheated and wrist pins broke. As these problems arose, one of the cars would be dispatched to Gallup or Farmington to pick up replacement parts and within a few hours things would be rolling again.

Dirt and dust troubled men and equipment. The cat operators had to wear goggles for protection from the dust which rose in thick, white clouds. The cars and trucks became stuck time after time and had to be pulled out of the powdery sea by the half-track.

At mealtime, the men brewed coffee over a campfire, taking a few minutes of welcome relaxation from the hard, grinding work of keeping the equipment moving.

Coordination between the front tractor and rear two tractors was essential but difficult. The men behind couldn't see ahead except by leaning far to one side. If the front tractor needed to stop, it was necessary to use hand signals — sometimes by relay — to get the rear cats to stop shoving. Going downhill, the rear tractors winched the heavy steel vessels to keep from rolling out of control.

The San Juan River, running full and fast, bedeviled the pipeliners. With trucks pulling and 'dozers pushing, the huge vessels edged into the stream. Stout cable from truck winches on the north shore slowly and carefully pulled their tremendous loads across the stream. Once they arrived at the San Juan Plant site, however, it was short work erecting the towers on already pre-

TOO HEAVY to be hauled on the highway from the railhead at Gallup,
New Mexico to Farmington, 180,000 pound loads like these were moved across
the desert at about one mile an hour.

PIPELINERS Buck Brewer and Norris Stogner brew coffee over campfire—welcome relaxation from the grinding work of keeping equipment moving.

CROSSING the San Juan River, running full and fast, took skill and coordination by El Paso's crews.

pared foundations. It took only about an hour to raise a sixty-seven ton steel mammoth into place, but it took three days to set up the big gin poles by which it was lifted.

At times during the slow struggle of moving the caravans from Gallup to Farmington, the crews reached a deadlock with Mother Nature. The weather conspired to bring winds resulting in even more clouds of choking dust which completely stopped forward movement. Then rains followed and softened the ground into a sea of mud. The big trailer sank into the mud, delaying the trip until the sand dried out in a day or so. In short, to quote one of Murphy's Laws: "Mother Nature is a bitch."

Undoubtedly, the most appropriate of Mr. Murphy's Laws slugged the San Juan Project at this same time with the force of a tidal wave. This is the principle that says: "In any field of scientific endeavor, anything that can go wrong, will go wrong — and at the worst possible time." By March, 1951, all but thirty-two miles of the San Juan pipeline was completed and ready to begin deliveries of gas, or so everybody thought until the U.S. Department of the Interior, in April, 1951, explained a new regulation. Secretary of the Interior, Oscar Chapman, issued an order that since El Paso's right of way went through Federal lands, the last sixteen miles of right of way through northern Arizona would not be granted unless the Company met certain conditions: namely, that the pipeline concede to operate as a "common carrier," available to transport gas from any other producer along the entire route. It would

work much like a railroad.

To Company President Paul Kayser, this meant that the Interstate Commerce Commission would get complete power to determine rates, regulations and rules in handling natural gas, making El Paso subject to dual regulation of the FPC and ICC. Their latest proposal would also mean that the Department of the Interior would have to pass on all gas matters before they were even submitted to the Federal Power Commission. The peppery Kayser called the regulation unacceptable and decided an injunction suit would be needed to restrain enforcement of the new regulations.

In May, 1951, the Company halted all construction on the line. Everything stopped —just inches from the goal line.

El Paso needed some top representation in Washington, D.C., and the law firm of Hogan and Hartson seemed to be the strongest for this type of action. Certainly the firm had the proper credentials. Frank J. Hogan was the celebrated trial lawyer whose skill resulted in a bizarre episode in the notorious "Teapot Dome" scandal during the Harding Administration of the 1920s. In this litigation, Albert Fall, Secretary of the Interior, was charged with issuing valuable oil leases on government land in return for a bribe allegedly given to him by Henry L. Doheny, a prominent oil man of great wealth.

The two men were tried separately. Fall had already been convicted when Doheny's case was called for trial. Frank Hogan's masterful defense secured Doheny's acquittal with the astonishing result, incomprehensible as it may seem, that Fall was convicted of accepting a bribe which was never given. (The white-columned mansion, once the home of the Albert Fall family, now ages ungracefully among the weeds at 1725 Arizona Street in El Paso, Texas.)

In 1951, the most highly respected trial lawyer in Washington, a member of the law firm of Hogan and Hartson, was Howard Boyd. He handled more cases in Washington than any other lawyer, so the Company retained Hogan and Hartson with the understanding that Boyd would handle its case. To lose it would have meant nothing short of disaster for the Company, as El Paso had already spent some $40 million on the pipeline and there was now no way to deliver the gas without going through a few precious miles of Federal lands.

Howard Boyd remembers the situation very well: "There were some rather amusing incidents that display the character of Paul Kayser whom I encountered for the first time in my life. I explained to him that in such a suit the government would have ninety days in which to respond. Mr. Kayser's immediate reaction to that was, 'We can't wait that long, and federal rules or not, we just have to have some solution to this problem in far less than three months.'

"I also pointed out to him that not only did the government have ninety days in which to respond, but at that season of the year the United States District Court often took a recess about June 15th and didn't reconvene until October. Here again, Mr. Kayser just swept that aside and said that he would expect a good lawyer to find the

Howard Boyd

solution within the short period of time.

"So," says Boyd, "I undertook to prepare the suit against the Secretary of the Interior. Because of my prior association with the Department of Justice I went down and talked to responsible officials explaining the need for urgency with the hope that they would cooperate with me in getting an early trial. And to the credit of the Justice Department I must say that they were highly cooperative."

Boyd immediately filed an application for injunction in the Federal District Court in Washington. And because of the high-priority time element to complete the pipeline, he obtained a hearing very quickly. It was assigned to Federal Judge T. Allen Goldsborough.

Goldsborough is remembered by many for the fearless action taken by him during World War II against John L. Lewis and the United Mine Workers. Lewis, with his threatening eyebrows, was without question one of the most powerful and feared labor leaders in the United States—the unchallenged and complete boss of the United Mine Workers. Lewis had called them out on strike at a time critically injurious to the war effort, and President Roosevelt was unable to deal with the situation.

However, the government filed suit against the UMW, and the case came before Judge Goldsborough. First Goldsborough ordered the mine workers to return to the pits promptly. When this command was ignored, the judge immediately slapped fines on the UMW of $3.5 million and on John L. Lewis personally in the

BY 1954, the Company was serving northern California through its San Juan pipeline (completed in 1951) from near Farmington, New Mexico, across northern Arizona to the Colorado River. Also in operation was the Permian-San Juan crossover line which ran diagonally across New Mexico. Other extensions and field lines had been added to the system by 1954.

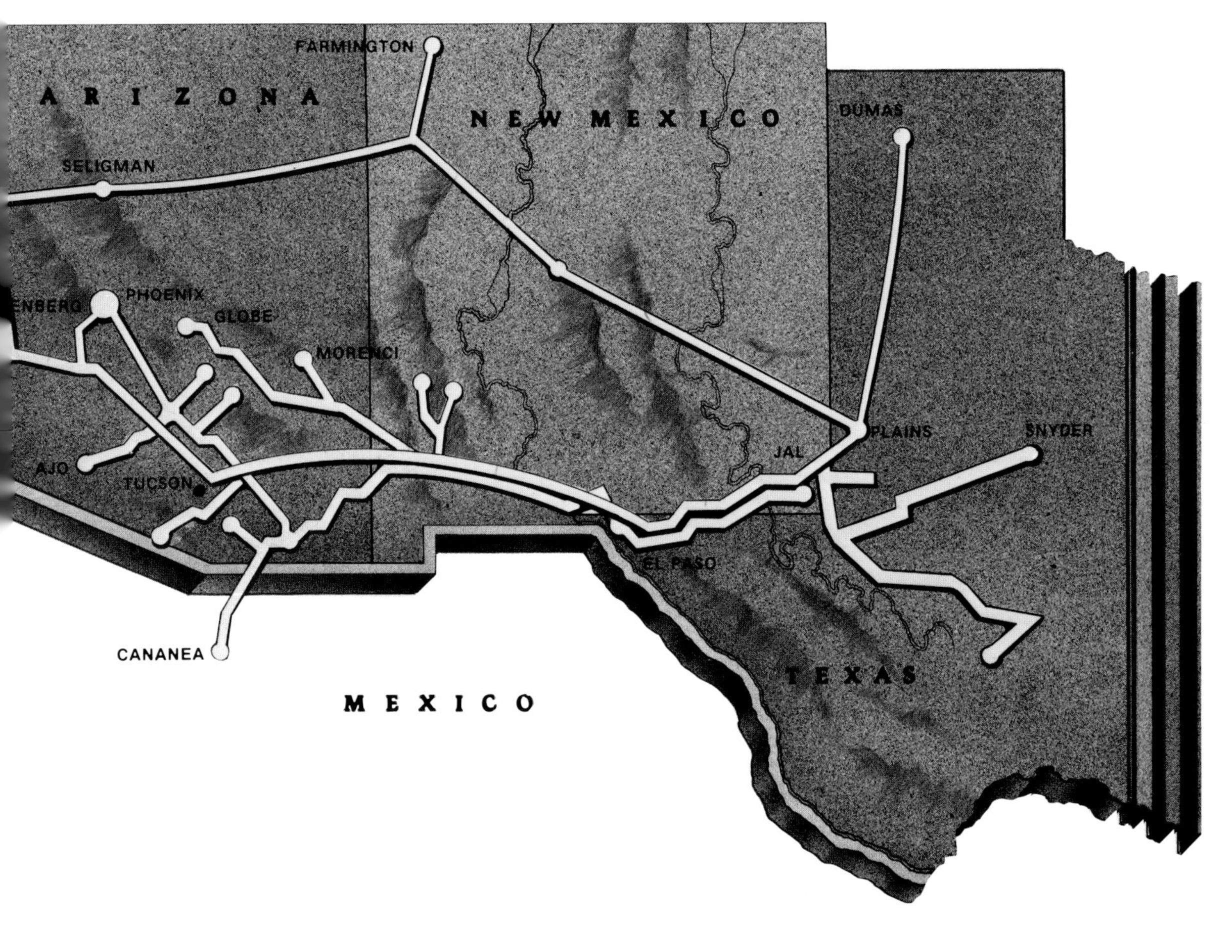

ARIZONA
NEW MEXICO
FARMINGTON
DUMAS
SELIGMAN
ENBERG
PHOENIX
GLOBE
MORENCI
PLAINS
SNYDER
JAL
AJO
TUCSON
EL PASO
CANANEA
MEXICO
TEXAS

amount of $10,000. Goldsborough went even further and said that, if necessary, the bank accounts of the UMW would be seized to assure payment of the fines.

Without funds to pay the miners while out of work, Lewis realized that the men would not support him, and for one of the few times in his life, he was forced to admit defeat. Lewis ultimately ordered the men back to work, giving as his reason that the war effort commanded top priority.

Meanwhile, in the El Paso case, after unending days of examining witnesses and documents, and after the testimony was in, Judge Goldsborough adjourned the case over one weekend to be reconvened the following Monday. In light of the testimony, Boyd and El Paso were confident that the decision would be favorable to the Company. Goldsborough had shown remarkable understanding of the facts and the necessity for quick and decisive action.

Then disaster struck again. On Saturday night, Judge Goldsborough died. And with his untimely death Boyd's hopes of getting a quick, favorable decision appeared shattered.

Boyd recalls that, "At the time of Judge Goldsborough's death, the U.S. District Court was about to go into recess for the summer months. One of the problems that confronted us was whether we could prevail upon a judge to undertake this complicated and important case which could infringe on his summer vacation.

"In the hope of making the assignment as palatable as possible to one of the judges, I proposed to the Department of Justice (which was representing the Secretary of the Interior) that they submit the case to a federal judge willing to take it on the record which had already been made. The Justice Department had already shown a cooperative attitude toward the case, and they agreed to my proposal—although at the time I suspected that they were reasonably confident that a judge couldn't be secured so late in the season.

"Fortunately, by pointing out the dire consequences facing El Paso from any delays in the case, one of the federal judges agreed to consider it on the record which had been made before Judge Goldsborough, thereby avoiding taking evidence anew— and permitting a prompt decision."

In June, 1951, Secretary of the Interior Chapman was ordered to issue right of way to El Paso to finish the San Juan Project. The judge said that El Paso need not sign Chapman's proposed agreement since the Company had built the line under a previous agreement and the Secretary could not insist on another. However, Chapman did not issue the right of way permit. "This," says Boyd, "gave rise to a very troublesome question of how El Paso should deal with the issue. Consideration was given to asking the Court to hold the Secretary in contempt. On this question, the Secretary would have been entitled to a trial from which he could have taken an appeal.

"All of this would have taken more time than El Paso could stand. The decision was made, therefore, that since the Court had ordered the issuance of rights of way, we were entitled to proceed as though the Secretary had done what the Court had ordered him to do.

C. L. PERKINS (right)
explains pipeline
construction duties to
Navajo crews through
interpreter.

GEOLOGISTS work
at the base of Shiprock,
famous landmark in
northwestern New Mexico.

HEAT EXCHANGERS cool natural gas after it has been compressed.

"Without further ado, orders went out to the spread laying the pipeline to proceed across the sixteen miles as rapidly as possible. The Secretary ultimately took an appeal to the U.S. Court of Appeals in Washington. But when the case was heard, gas was already flowing to California. El Paso's position was thereby strengthened by the fact that a reversal of the lower court would have entailed an interruption of the gas supply."

After his successful handling of this critical case, Howard Boyd was destined to go on to greater things.

"This, of course," says Boyd, "proved to be a tremendous milestone in my personal life, because following that successful litigation Mr. Kayser requested that I consider association with El Paso. Here again, I learned that Mr. Kayser did not readily accept a negative answer. But I told him that my roots were too deep in the District of Columbia. I told him that my father had been a practicing lawyer in the District before me, that I had been born and reared in the District of Columbia, gone to school there, practiced law there, I was identified with what I considered to be the best law firm and that I viewed my situation as having a possible bright future. And for those reasons, as much as I admired El Paso and its executives that I had met, I was content to stay where I was.

"As I mentioned, Mr. Kayser didn't then and doesn't still accept no for an answer. He persisted in his suggestions that I at least favorably consider association with El Paso. When I continued to decline these offers, he asked me whether or not I had ever been to Texas, and I allowed as how I had never been to Texas. He then said he didn't understand how I could make an intelligent decision about an important matter that might affect my future if I had never even been to the state. Actually, I did feel that perhaps to do justice to my own personal situation I should explore the invitation of Mr. Kayser further. And I would be less than frank if I didn't disclose in the meantime that I had developed a very strong personal affection for Mr. Kayser.

"I was very impressed with his statesmanlike approach to problems, his obvious business acumen, and I had convinced myself that association with El Paso would be an association with a company that was destined for success and promise. So, in response to the invitation to visit Texas, my wife, Lou, and I came to Houston and were met at the airport by Mr. Kayser. We spent about three days in the area meeting people who were prominent in the business—that included, among others, meeting Frank Liddel who was then General Counsel for the Company and who had been a law partner of Mr. Kayser.

"And I met Mr. Jesse Jones, who had been influential during the days of the Depression in helping financing the expansion of some of the pipelines. I also met any number of Mr. Kayser's close business associates as well as some of his personal friends.

"I'd had no experience whatsoever in flying on private planes, and when we had visited Houston to the point where I had some feel of the community, Mr.

INDIAN COUPLE is framed
against the Southwestern
sky at El Paso's Lindreth
Station in the productive San
Juan Basin of New Mexico

GLEAMING TOWERS at
Chaco Plant in northwestern
New Mexico.

PUMP HOUSE at the
gasoline plant of San Juan
River Plant near Farmington
is a maze of intricate piping.

SAN JUAN RIVER PLANT
treats gas before it moves
into the mainlines.

Kayser suggested that we fly to El Paso. This was to be the first occasion that I would have ridden on an executive airplane. We went out to the airport, and there was a Twin Beech there piloted by Guy Johnsen and co-piloted by Tommy Love, two chaps with whom I ultimately became acquainted and for whom I have the highest regard.

"We took off from Houston and had been out of the city for only a short period when we flew over an impenetrable cloud cover. You couldn't see the ground and there was no visibility whatsoever, when one engine coughed and conked out and there was a great deal of flurry in the cockpit. Before that situation could be dealt with the second engine began to cough, and I was sure doom was upon us. About that time the first engine caught on and everything smoothed out. It turned out, much to the embarrassment of the pilots, they had failed to shift from one gasoline tank to another and the engine performance was simply an indication that the tank on which they were relying had been exhausted.

"I suspect that since that initial ride in the Twin Beech I have undoubtedly flown several millions of miles in the Company's aircraft. And of course, it was a great tragedy when Tommy Love met an untimely death by drowning [while on vacation in Mexico].

"We went out to El Paso, and I was surprised as I'm sure many people are, to observe the radical difference in the vegetation there than what I had observed in the Houston area — the very dry climate that's in such sharp contrast to the humid climate that we're accustomed to in Houston and Washington.

"We met, of course, the executives in El Paso, visited some of the operations and stayed quite a few days. I came back with a fine impression of the Company, and not to my regret, I decided to throw in my lot with El Paso."

In 1952, Boyd joined El Paso Natural as a Vice President of the Company. By 1960, he was named President, replacing Paul Kayser, who became Chairman of the Board of Directors. Subsequently, in 1965, when Kayser stepped down, Howard Boyd took over the reigns as Chairman of the Board and Chief Executive Officer, destined to lead El Paso through some of the stormiest seas of its corporate life.

There was no question, however, that Boyd had the capability to handle the rough weather ahead. He is a formidable, unflappable spokesman in any group, ranging from an informal meeting to a dignified appearance before the Supreme Court. His staff kept him up to date on even the tiniest details, and his experience as a trial lawyer in Washington served him well.

By late summer in 1951, the San Juan Project neared completion. The San Juan River Plant was on stream and construction was finished at Navajo Compressor Station inside the Navajo Reservation near Klag-eh-toh, Arizona. Its completion marked one of the most unusual events in the Company's history. A large number of Navajos who helped build the station stayed

GRAND OPENING of the San Juan project was celebrated by a Navajo
parade and rodeo in the summer of 1951.

on as permanent operating employees. In
July, the Indians celebrated the grand opening with a barbecue and rodeo to which
Company employees and officials were invited as guests of honor. Indian dignitaries
and rodeo performers paraded before the
rodeo, and mountains of barbecue were
served to more than 2,000 Navajos, including members of the Tribal Council and several hundred El Paso Natural employees—
an example of appreciation and friendship
rare in the history of relations between
white man and Indian. During the ceremonies, the Navajo people presented the
Company three hand woven rugs bearing
the Company emblem as a token of friendship.

By early fall in 1951, the San Juan Division was on an operating basis. Service to
communities in northern Arizona had been
under way since mid-summer. In August,
natural gas from the San Juan Basin began
flowing to Pacific Gas and Electric Company for service to residents and industries
in San Francisco and northern California.

The San Juan Project had been a tough,
but highly successful undertaking.

R. W. HARRIS opens soda pop in chow line
where mountains of barbecue were served
to more than 2,000 Navajos and several hundred
El Paso Natural Gas employees.

173

FARMINGTON:

Transformed by natural gas into one

of New Mexico's most important cities

BACK IN 1951, when El Paso Natural was building its San Juan pipeline to gather, treat and transport gas from the San Juan Basin to California, the optimistic *Farmington Daily Times* reported cheerfully on the growth of the little town of 3,000 people. "It is generally believed," said the *Times*, "that Farmington is likely to reach around 10,000 in population before there is any leveling off in the community growth. The sleepy little village of yore is giving way to a young and virile metropolitan community."

This was probably the greatest understatement since Noah looked up at the sky and said, "It looks like rain."

When El Paso opened up the abundant new markets for natural gas, major oil and gas companies began moving in. And right on their heels were supply companies, truckers, drilling mud companies, geologists, machine shops and other service industries. It was like a giant chain reaction. Trailer camps mushroomed on the edge of town. All these people, many of them from the oil fields of West Texas, New Mexico, Oklahoma and Kansas, needed domestic services. Motels, cafes and clothing stores opened their doors to the newcomers.

Although it was a friendly invasion, much of the time Main Street in Farmington looked like D-Day in Normandy. Noisy, mammoth trailers hauled drilling pipe, and heavily loaded service trucks strained bumper to bumper up the street bearing insignias on their doors representing firms from Houston, Odessa, Tulsa, Midland and dozens of other oil towns. Texas license plates began to outnumber New Mexico's as this whole new breed of people changed the complexion of Farmington. Even the speech patterns of northwestern New Mexico gave way to the unmistakable drawl of transplanted Texans.

The town raced to keep ahead of the demands of the invading hordes. One businessman commented that it was the greatest thing since packaged tortillas. City planners and utility companies kept busy revising their estimates of the town's growth. The Farmington boom was unique. It was entirely different from the frantic Texas oil booms of the Twenties. By 1960, Farmington's population reached 25,000.

In some of the earlier booms in the Permian Basin, whole towns appeared almost overnight on the raw prairie. Drilling crews and service outfits lived in tents and flimsy wooden shacks. Whole blocks of buildings were hastily constructed of corrugated sheet metal and wall board. Many of the new citizens were merely opportunists who rushed in to get their share and then pulled out just as hurriedly to try their luck on another, newer boom camp.

But not so in Farmington. Here the picture was so different it was sometimes hard to believe that the town was the center of the great San Juan oil and gas province. For one thing, drilling methods had changed dramatically during the past twenty-five years. Instead of a countryside dotted with hundreds of skeletal derricks as in the Permian Basin or California or East Texas, drilling rigs were portable and were moved off the site when a well was completed. So the casual traveler through the San Juan Basin saw only an occasional Christmas tree valve marking the site of a producing well. Luckily for Farmington, its buildings were substantially constructed right from the first. There would be no nest of wooden shacks to clean out in years to come, because neat trailer courts housed the hundreds of families who were not able to find immediate housing in town. Most of Farmington's new buildings were made with brick, steel and masonry, comparing architecturally with some of the best in the country.

Although the peach and apple orchards were being pushed back to make room for new housing subdivisions, some builders managed to keep a few of the trees on each new lot. It wasn't unusual to see people moving into a brand new home complete with several huge old apple and peach trees built right into the landscape. Piñon-dotted mesas also provided ample scenery from the picture windows of Farmington's new ranch-style homes.

The boom in Farmington in the 1950s was certainly its biggest, but history shows that people had been coming to this shady oasis for a long time. In the late 1700s, Spanish padres passed this way looking for a route to connect Santa Fe, New Mexico with Monterey, California. American trappers, as early as 1822 wintered along the San Juan River near present-day Farmington. The first permanent settlers were prospectors out of the Colorado mountains and stockmen coming down to graze their stock on the grass covered mesas and valleys of the San Juan.

Before 1876, the Farmington area was largely a part of the Jicarilla Apache Indian Reservation. But on July 4 of that year, the U.S. government opened the land to white settlement and the first few families drove in with their covered wagons. Subsequently, the lands were made available for homesteading. The town, twenty-three miles south of the Colorado line, was named by Milt Virdin, one of the first settlers. He helped Judge S. D. Webster, L. C. Coe, and a man named Boran lay out the townsite in 1879. More stockmen began to arrive, as it was a natural place to buy green vegetables and forage for their animals, and they called it the Farming Town. Virdin suggested dropping the "w" and combining the two words into "Farmington." Everybody agreed that this was a grand idea, so the town and post office took on the new name.

By 1881, the settlements of Farmington, Aztec, Blanco and Fruitland attracted more homesteaders, and the Indians, quite naturally, resented this intrusion by the white man. Early Farmington pioneers had more than their share of trouble, grimly holding on to the newly-acquired lands.

But long before the Spaniards and the Yankees and the Apaches and the Navajos and Utes came on the scene, the San Juan Valley supported another civilization, that of the Pueblos. Some years ago while working on West Main Street in Farmington, highway builders unearthed remains of a primitive pit house. The owner of this pit house was undoubtedly one of the Basketmakers who, archeologists say, lived in the area almost 2,000 years ago. Other ruins in and around Farmington and stretching for miles along the lush river valleys are those left by the apartment building Pueblos of a later age.

The valleys of the San Juan River form an area in which the Pueblo Indians reached a high stage of civilization. By about 600 A.D. they had settled in villages in the area, and as their communities grew, the people improved their arts and crafts and methods of farming. By 1100 the Pueblos had reached a high degree of achievement and it was about this time that the great pueblo at Aztec (fifteen miles east of Farmington) was constructed. It was once a building of 500 rooms, three stories high. The name "Aztec" was applied to the ruins by the first white settlers who had a mistaken idea of the ancient builders. Actually, these Pueblo Indian builders were not related to the Aztecs of Mexico, but apparently a lot of people thought so at the time. Aztec Pueblo flourished for a time and then died. The entire Basin was abandoned in 1300, probably due to a massive twenty-five-year drought which affected a large part of the Southwest.

Drought, of course, is always a threat and Farmington's nine inches of annual rainfall still creates a delicate balance between man and his ability to survive in the arid Southwest.

Many years later — perhaps a hundred years before Columbus discovered the new world — the present-day Indians arrived. These were chiefly the Navajos who have inhabited the area for about 600 years. They named the spot Totah, which means roughly "where the three rivers meet." Converging on present day Farmington, water from the San Juan, La Plata and the Animas rivers still plays a big part in the life of the area. These rivers irrigate fruit orchards and more than 100,000 forty-pound boxes of San Juan Basin apples are shipped every year from processing plants in Farmington.

Bill Parrish, superintendent of El Paso's San Juan Division, will tell you

FARMINGTON, in far northwestern New Mexico was
transformed by natural gas into one of the state's major cities
by the mid-1950s.

that natural gas — and the possibilty of developing energy from the vast coal
fields nearby — are the reasons for Farmington's continued steady growth.

"In spite of the gas shortage," drawls Parrish, (who has the reputation of
being the only man in the Company who takes three syllables to say the word
"oil") "we're managing to hold our own up here. We still have a drilling program going on, although it's not quite as large as was originally contemplated.
But in 1976, El Paso tied in 166 new wells in the Basin during the year. This is
called infill drilling, and what that means is that we got permission from the New
Mexico Conservation Commission in 1974 to change the spacing on Mesaverde

wells from 320 acres to a spacing of 160 acres. That allows us to go in and drill two additional wells in each section.

"The Mesaverde formation is about a mile deep, but it will vary according to what part of the Basin it's in. We also get production from the Pictured Cliffs formation (which is shallow — about a half-mile deep) and the Dakota (which is a mile and a half or two miles deep). But the bulk of the gas that we get in the San Juan Division comes from the Mesaverde.

"To maintain our level of deliverability out of the basin to meet market requirements, we've got to drill additional wells each year. And, so far, we've been maintaining, even though as time goes on, a gas well will produce less and less due to the decline in the reservoir pressure. Every year some wells are just plugged off because the gas underground is exhausted."

Parrish's division employs about a thousand people, involved in every phase of the operation — including maintaining some 5,000 miles of roads.

San Juan winters can be pretty unreasonable, "But," says Parrish, "we try to keep our roads passable all year around. We have to get to the wells to check our dehydration and metering equipment, especially in the wintertime when we need the gas. Actually, we're better off if the roads are frozen. Where we really get into problems is in the fall when it starts raining or snowing, and maybe it'll freeze at night, but it will thaw in the daytime, and then we don't have any frozen base under us. The same thing happens in the spring, and when that frozen base thaws, the mud underneath is sort of bottomless. That's when we have to rely on our "mudbuggies" because they're the only things that'll move in conditions like that. Usually, though, the mudbuggies are used to retrieve equipment that breaks down or is hopelessly stuck in deep mud or snow."

Richard Butchofsky, who heads up El Paso Natural's Transportation and Purchasing Department, must be given full credit for San Juan Division's mud-buggies. "I bought that first mudbuggy right here in El Paso from an ol' boy who built it to go hunting in Alaska," says Butchofsky. "It had big airplane tires on it. R.W. Harris let me buy it and I took it to Farmington to try it out. Leonard Fincher and I drove clear out to Ojito, where the swamps were. We were goin' like mad and I said, 'Leonard, we oughta' be gettin' to where we can test this pretty soon. We're already in water and grass.' He said, 'Well, I guess we ought to.' So I said, 'Let's stop a minute, I'm gettin' kinda' shaky up here on this damned thing.' So, we stopped and I stepped off and went half way up to my waist in water. We had already tested it! That's when we started building them for the different districts. We have about seven now, and they'll go anywhere."

Parrish recalls his first days in Farmington in 1950 with a seismograph crew doing exploration work: "Main Street was only two lanes wide; that was the highway. If you wanted to park downtown by the sidewalk, you had to park in dirt, and that's the reason why in the spring or fall you could get stuck up to the doorhandles." Parrish started to work for the Gas Company in 1952 as a junior engineer and has lived there ever since. "When I first went to work for the Company," says Parrish, "R.W. Harris was in charge here. And following Mr. Harris, Jack Stricklin came. Following Stricklin was Ed Alsup, then H. P. Logan and Thurman Bailey. After that E.W. Woody ran the division, and when he retired several years ago, I took over. So I've had a lot of good teachers ahead of me.

"Farmington is a nice place to live. I didn't like it years ago when I first came up here, but you stay here awhile and it grows on you. The winters are not too severe. About every third year we'll have a real cold one, but normally they're not all that cold. And nowadays, the city continues to grow and is large enough that you can get most of what you want and what you can afford. I'd say Farmington has finally "arrived"; we now have a McDonald's hamburger stand!"

Natural gas is still the biggest thing in Farmington, with 10,700 gas wells producing in the northwest part of New Mexico. There's still a tinge of excitement in the air, the restless feeling that comes from moving and growing. But there's also a feeling of well-being and leisure. With all its activity, here is still a town in the real tradition of the West. It remains an outpost on the edge of the vast Navajo Reservation, and there are more than 4,000 Indians living in Farmington. The total number of Navajos in the tribe is estimated at 140,000 and about 40,000 of these trade regularly or occasionally in the city. On Farmington's busy street you see Navajos in their colorful turquoise and silver. You see ranchers and geologists and even white-shirted businessmen. And mingled throughout, the unmistakable khakis and hard hats of the gas field hand. These people and the things they represent are the real personality of this growing city, and most of them will tell you in no uncertain terms that "life is good in Farmington."

THE SAN JUAN PIPELINE

Selecting the route

The old axiom that a straight line is the shortest distance between two points is not necessarily true when it comes to building a pipeline. And the San Juan pipeline was no exception.

Civil engineers and right of way men met the usual obstacles crossing some 226 miles of the vast Navajo Indian Reservation — as well as lots of new ones; heavily forested areas, pre-Columbian villages, and many people who had never heard of a natural gas pipeline, much less the prospects of having one built across their lands. The route had to be located to make use of favorable terrain, and for that reason, pipelines which look bullet-straight on maps really contain many gentle curves.

The San Juan line took many months of designing, planning and negotiating approval from landowners before the first pioneering bulldozer could start clearing the right of way. On the Reservation, part of the land is owned by the Tribe, part of it is called "allotted" land and is owned by individual Navajos.

Following their usual procedure, El Paso's civil engineers went on the Reservation to survey the possible pipeline route, and a right of way man was with them all the way. He went with the survey party and informed property owners and occupants on tracts of land a mile on each side of the projected route that a survey party was on the way and would possibly survey a route across their property. The agent explained (in most cases with the aid of an interpreter) that rights would be negotiated at a later date if the survey line was to run through the owner's property. When the survey was concluded and maps prepared from the field notes, each individual Indian whose allotment was crossed was located and his signature or thumb print obtained.

Bill Howard, who heads up El Paso's Right of Way Department, explains that buying easements through the Reservation was a new experience for him, but that in most respects it was handled in pretty much the same way it's being done today. "Usually, right of way is bought by the rod — that's sixteen and a half feet," says Howard. "This goes back to the early days of the oil companies. Why they started buying right of way by the rod I don't know. We've tried to

RIGHT OF WAY MAN gains permission from a Navajo
family to purchase an easement permitting pipeline installation
and maintenance on the reservation.

buy it by the acre, but people always fall back on a linear measure for some
reason. In every case we attempt to get at least sixty-foot easements with each
rod."

One of the big problems in purchasing right of way is finding the owners of
the land. "But," says Howard, "we've had a lot of experience in detective work
and so far I don't know of a single landowner who we haven't been able to track
down somewhere." Between El Paso's gathering systems and the California border, Howard estimates his group had to find and make contracts with some 400
individual Indian landowners. This didn't include federal, forest and privately
owned tracts of land in Arizona and New Mexico.

The Texas Panhandle was another area that took some real digging to find
the owners. "In the early days, the early 1900s," says Howard, "they had these

land sales out of Chicago. They ran trains out of the Midwest bringing people down to spend the week in Amarillo and Lubbock and those areas. So some of the big ranches were split up for these sales and there was a lot of land owned in 160-acre tracts. When we went through the big middle of that in the late Forties I chased people all over Minnesota, Wisconsin and even got over into Michigan for pipeline right of way that we had to have."

What happens if a landowner refuses to grant right of way through his land? "Well," says Howard, "in 1949, all interstate natural gas pipeline companies were given the right of condemnation by Congress because the pipelines are designed to be in the public interest. Naturally, we've had our share of problems with some people, but they know we can always go to the courthouse and get some kind of a settlement. In the past, most people were glad to have the extra money we paid for right of way. But sometimes we ran into a brick wall, even though we tried to be fair and equitable with everyone. Some big ranchers refused us, but we just sat down with them until they got so tired of looking at us they said okay.

"We've even had a gun pointed at us occasionally. One time we had a gentleman who we thought we had already made a deal with, and he came out with a shotgun and ran the survey crew off. I went out there and had a talk with him and he kept that shotgun in his hand all the time. I finally told him that we could talk just as well if he could put that shotgun down."

Even Paul Kayser tells of a similar incident he was involved in during the early days of the Company. "I'll never forget one instance when a woman stood at her fence where we were coming through with a pipeline. She had a 30-30 pointed at A. L. Forbes and said: 'The first one of you sons of bitches that comes any further, I'm goin' to blow your head off.' That's what you call impasse.

"Well," says Kayser, "Forbes went a little closer and said, 'I just want to talk to you because I think we can satisfy you. She finally agreed to let him come up and talk. Of course, we had to make some concessions, but we got through. It was a dangerous situation because that woman would have shot somebody if we had come on in and started construction in spite of her."

In addition to paying for right of way, El Paso pays for crop loss or other damages that might have been caused during the construction. Even later, if damages are caused during maintenance or operation, they are corrected or paid for. The property owner reserves the right to full use of the land so long as it's consistent with the rights provided in the easement contract.

Howard remembers one incident of property damage that came close to being an impasse. "We had a river crossing one time near a lady's house that had a beautiful cedar tree in the yard. Our people had agreed that we would go around the tree with the pipeline. But when the contractor came in he said there was no way he could make the river crossing and save the tree. So they called me and asked me to come and help. The lady was crying and throwing a

CIVIL ENGINEERS survey proposed pipeline route.

fit, so I went in the house and sat down with her for about an hour or two, got her settled down and convinced her that we had to take the tree. I told her we'd buy her another cedar tree when we were through, but she said that the family just couldn't stay there while it was being done. So I said, 'Well, alright, if you want to go over to the beach and stay two or three days you can do it at our expense and we'll take the tree out while you're gone.' So they loaded up and headed for the beach. I think it cost us about $200, but we got the tree out of the way, planted another one, and got the pipe moving. And the family had a good time of it."

When the surveyors, right of way crews and construction spreads have completed their jobs, the project doesn't end there. A finished pipeline is out of sight, but never out of mind. After the earth has settled over the right of way, the pipe is zealously attended, cleaned on the inside, protected by electrolysis against corrosion and repaired when damaged. This work is done by trained crews who are assigned to pipeline districts established along El Paso Natural's entire system. Patrolling the rights of way at regular intervals gives an extra ounce of protection to the planned maintenance program. Patrols periodically cover every mile of El Paso's system—by truck, on foot, or from the air.

THE SAN JUAN BASIN

A record of the rocks

Millions of years ago, in what geologists called the Cretaceous Period, the San Juan Basin was part of a seaway that extended from the Gulf of Mexico northward through Alaska along the general path of the present-day Rocky Mountains. This is known as the Rocky Mountain trough or geosyncline. Through the ages layer after layer of sediment built up on the sea floor. Some of it came from the surrounding land masses as a result of weathering and erosion. Other sediments were deposited as a result of decaying plants and microscopic animals that had completed their life cycles. Shells of clams and oysters, skeletons of fish and sea mammals, seaweed, algae and all the other organic by-products of the abundant life of an ocean settled to the bottom of the sea.

Just as the oceans of the world today slowly change, the oceans of the prehistoric world constantly changed. The difference of a few degrees in temperature or fluctuation of water depth would cause the type of sea life to change. As part of the land mass of what is now North America began to rise, the various layers of sediment were pushed up and out of the water, revealing beds of limestone, layers of shale and sandstone. At some time part of the vast layers of rock and soil were folded and distorted. In the San Juan Basin they tended to take the shape of bowls, low in the center and high at the edges. Glaciers moved into the north part of the basin scrubbing off the topmost ledges and filling the valleys with rock debris. Although the surface of the San Juan Basin is still mountainous, it is not nearly as pronounced as it was before the erosion of time, wind, water and glaciation took its toll.

Within the basin today are the various sediments that accumulated at one time on the bottom of the sea. Along the margin of the basin these sedimentary layers crop out at the surface. A typical example is the Pictured Cliffs Formation which can be seen near Farmington, New Mexico. Some of these rocks have prehistoric Indian drawings painted on them — thus the name Pictured Cliffs.

PICTURED CLIFFS formation outcrops near Farmington, New Mexico. This gas bearing formation dips from the surface downward thousands of feet into the basin.

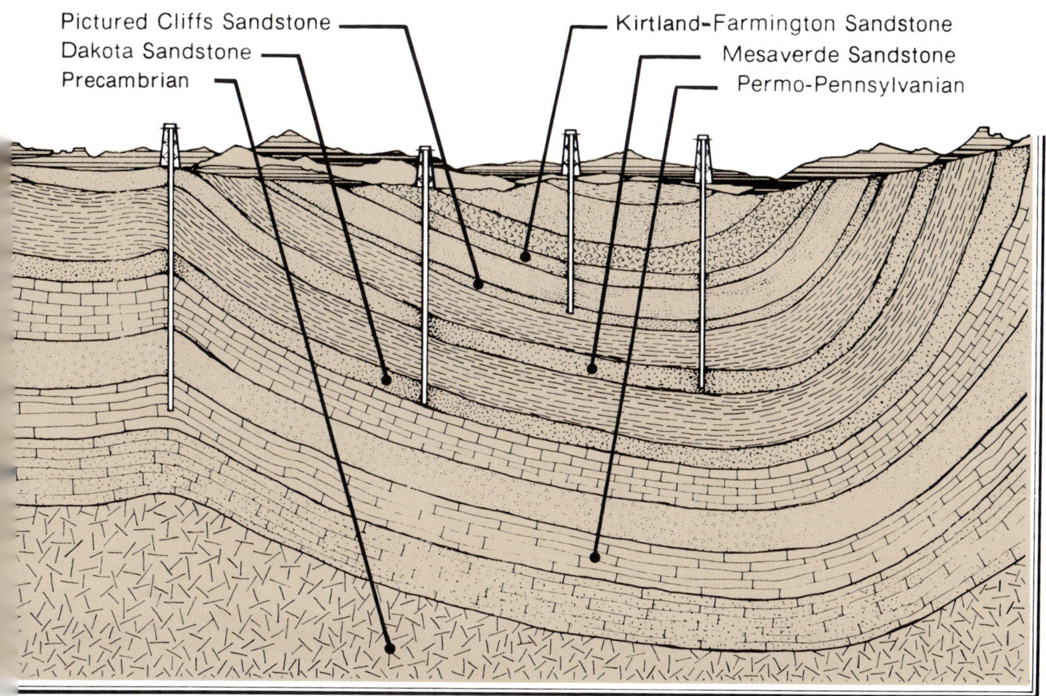

PRODUCING FORMATIONS are marked in this diagrammatic northwest to southeast cross section of the San Juan Basin.

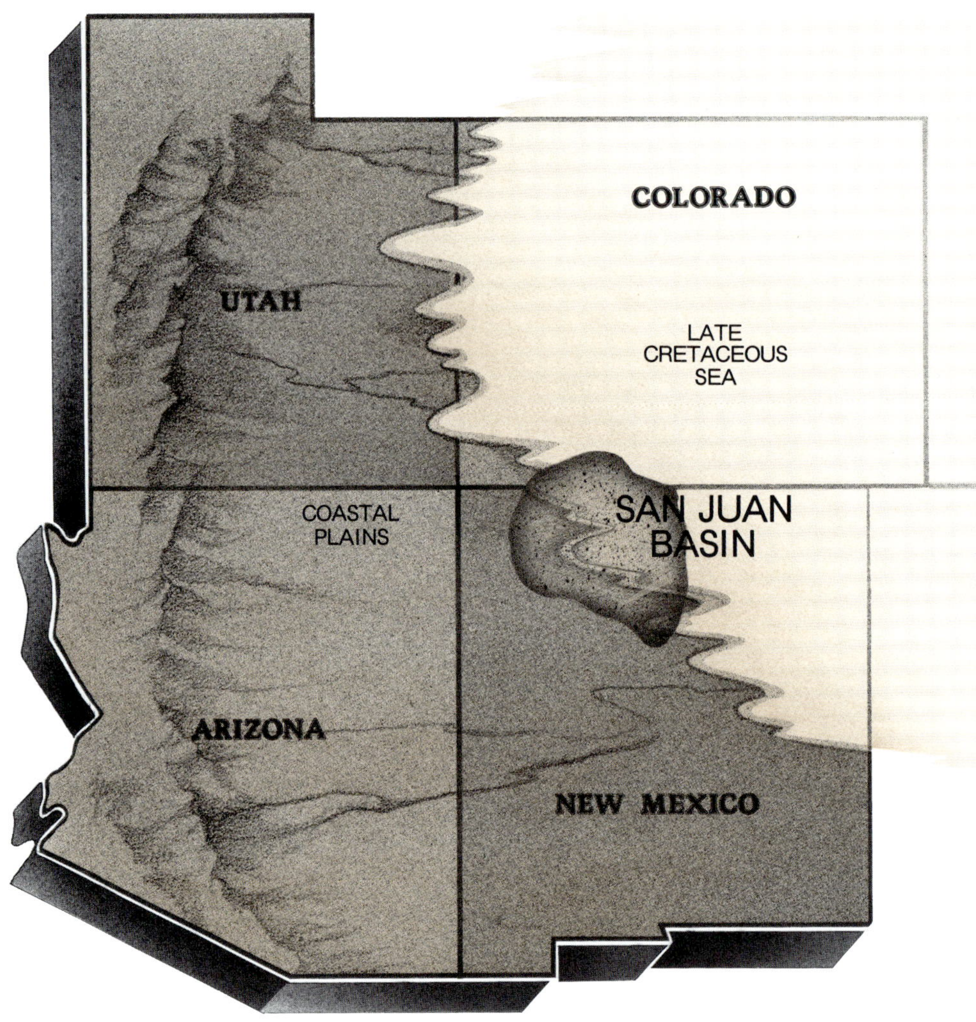

GENERALIZED location of Late Cretaceous Sea in the San Juan Basin region approximately eight million years ago. The shoreline was only a few miles east of present day Farmington, in northeastern New Mexico.

This gas-bearing formation dips from the surface downward thousands of feet into the basin.

In many cases, natural gas which was formed by the decomposition of animal and plant life, migrated into sandstone only to become trapped by a "roof" of solid rock immediately on top of the sandstone reservoir. In other cases the gas was generated in a layer of limestone and trapped there by super-compacted layers of the upper part of the formation. Consequently, gas is found only in strata of sedimentary rock. You might compare it to a blanket spread out horizontally, deep underground. If you move too far in any direction you may go beyond the horizontal limits of the "blanket" and thus get a dry hole, even if you tap a formation which might be producing gas a few miles east or west of your drilling rig. Likewise, the unknown target depth (the gas-bearing formation) must be determined by pin-point drilling accuracy. In short, the trick is to discover the reservoir and then how far, horizontally, the gas spreads within the particular stratum. Does it extend one location away? One mile? Or twenty? Or eighty?

Natural gas is where you find it, and the only way to know is to drill for it.

EL PASO NATURAL GAS

9

THE FIFTIES

Happening Years

Across the length and breadth of America the decade of the Fifties began with a feeling of doomsday in the air. Americans became awesomely aware that Russia had detonated its first atomic bomb, and the public reacted with a rash of bomb shelter building. U. S. scientists quickly went ahead with plans to build the biggest bomb of all—the H bomb. On November 1, 1952, on a small Pacific atoll named Eniwetok in the Marshall Islands, the U.S. created the largest explosion ever witnessed by man. But in spite of world foreboding, the country trudged ahead as the Korean War still made the headlines. President Truman fired General MacArthur in Korea after 200,000 Chinese and an unpublicized number of Russian technicians moved across the Yalu River.

Americans cheered the 1952 election of Dwight Eisenhower, who won by a landslide after an estimated seventy million people saw the Democratic and Republican national conventions on TV. With the signing of the Korean peace agreement, people became less nervous about the possibility of another world conflict and settled down to a period of fun-loving prosperity the nation hadn't known for many years.

By the mid-Fifties, the mood of most Americans changed from gloom and doom to the new-found joys of living in an affluent world. People sat in front of their television sets to watch Milton Berle, "I Love Lucy," "The Honeymooners," "The Phil Silvers Show," Sid Caeser's satirical

"Your Show of Shows," "The Ed Sullivan Show," and other offerings that some alarmed preachers predicted would corrupt the morals of the young.

Oddly enough, as the country returned to normalcy, the Fifties became another age of crazy fads reminiscent of the Thirties. College campuses were swept by strange rites called "stuffing," which amounted to cramming as many kids as possible into small spaces. In California, forty students claimed a world record as they crammed themselves into one Volkswagen. Twenty-two students packed themselves into an outdoor phone booth. Texas was still the largest state in the union.

Plastic rings called Hula Hoops flooded the market and millions of kids all over the country were spinning them around their bodies in contortions that brought a windfall of new patients to orthopedic clinics.

The age of swing gave way to a wild new sound called rock and roll, a mixture of country-western and rhythm and blues. The term was coined by a New York disc jockey from an oldie called "My Baby Rocks Me with a Steady Roll." But it took the bumping and grinding gyrations of a twenty-one-year-old Memphis truck driver named Elvis Presley and his highly amplified guitar to turn rock and roll into a national phenomenon on the music scene for years to come.

On the silver screen, Marilyn Monroe became the sex symbol for a generation, and other names appeared in Hollywood: Marlon Brando, Charlton Heston, Rock Hud-

son were to become unforgettable male stars, and Bridgitte Bardot was being banned all over the country.

For El Paso Natural Gas, the decade of the Fifties became sort of a golden age. Demands for more and more gas zoomed to new heights. In June, 1952, the FPC gave its approval for El Paso to increase its deliveries to northern California by 150 million cubic feet per day. An increase to southern California was also authorized, as was delivery of twenty million cubic feet a day to southern Nevada. The Company acquired two thirds of the stock of Utah Natural Gas Company, and plans were laid for construction of an intrastate pipeline to carry gas from the Clear Creek Field to Provo and Salt Lake City.

On September 29, 1952, the desert air was still crisp and cool at seven in the morning. At one of the most remote spots along El Paso Natural's entire system, an intent group of men stood watching a large gas turbine being readied to go into operation.

The occasion was the opening of the Company's first gas turbine powered compressor station, located at a tiny spot on the map called Cornudas, seventy miles out in the desert east of El Paso. Preliminary sketches of Cornudas Station made their appearance on engineers' drawing boards almost three years before. Construction began early in 1952, after months of designing, planning, computing and changing. In spite of the simmering heat of

INTERIOR of a
turbine compressor
station shows gas turbine
engine which works
much like a jet
airplane engine.

TURBINE STATIONS,
in common with other
El Paso facilities are
carefully overhauled to
insure efficient operation.

191

summer, building went ahead right on schedule — doubly tough because this station was to be one of the first of its type ever constructed for use in the natural gas industry. One miscalculation could spell disaster later.

The group of men talked quietly as charts were given a last minute double check, figures were recomputed and instruments on the big panel were watched. A switch was thrown, and gas from the large twenty-six and thirty-inch southern pipelines rushed into the compressor. There was a moment of anxiety — then smiles all around. Everything was working smoothly and Cornudas Compressor Station, the first of ten stations of this type then under construction, was on the line.

As the 5,000-horsepower turbine at Cornudas began turning, El Paso made a pioneering step in the long distance transmission of natural gas — the nation's first large-scale application of a gas turbine driven centrifugal compressor. The logic applied to installation of this new machine resulted from plain old-fashioned economics. Gas pressure drops as it moves through a line, and to get the most in efficiency from hundreds of miles of large-diameter pipelines, pressures must be kept as high as possible. This means that the higher the pressure, the faster the gas moves to market. It also means that more gas is being pushed into the pipe.

There are several ways to increase the flow of natural gas to market through pipelines. In some cases, it's possible to add horsepower to already existing compression plants along the line; however, in 1952, El Paso's compressor plants were already operating at capacity.

An alternative — and a very costly one — is constructing a new pipeline alongside present lines to transport additional gas. After looking into all aspects of the problem the Company's engineers felt that the best method would be to install additional compressor stations along the already existing pipelines, particularly since the Korean War and its resulting steel shortage placed new pipe in short supply. New stations could boost gas pressures considerably and the existing pipelines could be used to a greater extent.

It was at this point that gas turbine driven compressors entered the picture. This type of unit has a low compression ratio — which means that it is capable of compressing "a lot of gas a little bit" — just what El Paso needed in its intermediate units. The new turbine stations would actually relieve the older reciprocating installations by taking off some of the load. It would decrease the amount of work normally required and place them in a position to pump greater amounts of gas with no extra effort.

"Before we got into this," says John Eichelmann, "we had no knowledge or experience regarding gas turbines. The whole thing was Perkins' idea. He asked Walt Miller and me to look into the turbine idea so we investigated first with General Electric. We needed seven turbines and G.E. told us they'd have to tool up for that size of an order, but we were convinced that gas driven turbine compressors would do the job on El Paso's system.

"After our investigations, Perk immediately decided to take the plunge, and we literally put G.E. into the gas turbine business. He just picked up the phone one day, called G.E. and increased the order to twenty-eight turbines. Just like that. Perk had faith in us, and all credit for the future success of the turbines should go to him. You have to remember that one turbine engine in the 1950s cost almost half a million dollars and he ordered twenty-eight of them on the phone. That's what you call sticking your neck out."

Until Cornudas Station went on-stream, reciprocating compressors did all the work. In this type of unit, a piston slides back and forth within a chamber pumping gas in much the same way a tire pump compresses air.

In the centrifugal compressors run by turbine engines, however, the impulse which moves the gas results from the rapid rotation of a bladed wheel. The principle of the gas turbine which powers this type of unit is the same as that of a jet engine. Gas burned within a chamber "jets" out of the chamber at high temperatures and at great speed. This swift, powerful stream of gas is directed against a turbine wheel on a rotating shaft. The force of the expansion of this gas rotates the turbine which in turn drives the compressor. Natural gas from the pipelines enters and is compressed by the centrifugal force exerted by rotation of the bladed wheel.

Before the coming of the new turbines, the Company's reciprocating stations were located about 100 miles apart. With the new expansion, two turbine stations would be constructed between every reciprocating installation. Thus, natural gas would be compressed about every thirty-three miles.

The early Fifties saw major changes in the management of El Paso Natural. A newly-created Operating Department was organized in El Paso: Estin Scearce, who served as superintendent of what was originally called the Jal Division, transferred to El Paso as operating manager of the Permian Division; Walt Miller, who headed up the Mainline Compressor Department, became operating manager of the Southern Division; D. H. Tucker, who was headquartered in Houston as a Vice President, was elected to the Board of Directors; John Eichelmann, formerly chief engineer, was made a member of the new Operating Department to coordinate all engineering matters; R. W. Harris, who ramrodded the building of the San Juan Division and remained temporarily as its superintendent, moved from Farmington to El Paso with the new title of operating manager of the San Juan Division; and Jack Stricklin, formerly transmission superintendent, took over the reins of the San Juan.

Stricklin remembers things moving quickly. "Perk called me in," says Stricklin, "and Steen was with him. It was the 27th of April, and he said if I wanted it, he'd send me to San Juan as division superintendent. I wanted it. A little more money and I could learn somethin'. And uh, he

R.W. Harris

Estin Scearce

said there was no real hurry about it but 'why don't you be there by May 1st?' That was the 27th of April, my wife wasn't told, my kids were in school and were gonna' be in school for another month. I had furniture to sell and a house to sell and another one to buy. But by May 1st, I was in Farmington."

In 1953, as El Paso reached its twenty-fifth year of operation, far from slowing down, the Company marked the event with ever-greater expansion efforts. Home Office employees could look out of their windows in downtown El Paso and see the eighteen-story steel skeleton of a new Home Office Building nearing completion. At the southeast corner of Texas and Stanton streets, a modern landmark was rising to dominate El Paso's skyline. Slated for completion in 1954, it would be the tallest building in town.

Twenty-five years earlier, it would have been difficult to imagine that the embryonic beginnings of El Paso Natural in one room of the First National Bank Building would grow to fill an eighteen-story building. Yet the phenomenon was taking place, just twenty-five years later and only two blocks down the street from the original one-room office. In 1953, Home Office employees were housed in ten different buildings in downtown El Paso.

In the spring of that year, 350 guests attended a party at El Paso Country Club honoring C. C. Cragin, who for twenty years of major accomplishments had served the Company as its Vice President and general manager, and who had resigned from the position. He would continue to serve the Company as a business and engineering consultant. Succeeding Cragin was C. L. Perkins, who was named general manager. H. F. Steen became general superintendent of El Paso Natural.

Meanwhile, pipelines continued to grow. A crew of Navajo Indians lowered in a section of thirty-inch pipe in an Arizona pine forest. Six hundred miles to the east, a pipe truck deposited its load of steel along the right of way near Midland, Texas. In New Mexico, a new turbine cranked up for operation on the southern mainline. In

Walt Miller

John Eichelmann

central Utah, crews made final checks on the new pipeline to furnish natural gas to Provo and Salt Lake City.

This hum of activity was all part of an El Paso expansion program that kicked off in 1953, the largest it had ever undertaken. At the completion of this expansion in the early Fifties, the Company would be moving two billion, two hundred thousand cubic feet of gas a day through its pipelines. This burgeoning growth was actually four different projects: the "400 Million Case," the "Exchange Agreement with Permian Basin Pipeline Company," the "Utah Pipeline Project" and the completion of many facilities which had been authorized by the FPC in 1951 and 1952.

THE 400 MILLION CASE

Largest of these projects, and the largest single application ever made by El Paso Natural until that time was the 400 Million Case. In June of 1953, the FPC granted permission for the Company to increase its

deliveries by 400 million cubic feet of gas a day.

Immediately after receiving permission from the Commission, El Paso started construction on facilities consisting of some 770 miles of thirty-inch main transmission line and about sixty miles of twenty-four-inch line together with related gathering lines, purification plants, gasoline extraction plants, compressor stations and other necessary facilities.

The new thirty-inch originated in the Permian Basin of West Texas and extended across New Mexico to a point near Gallup where it tied-in with the existing twenty-four-inch from the San Juan Basin. From that point west the new line paralleled the existing San Juan line to a point near Kingman, Arizona. In addition, sixty miles of twenty-four-inch was installed between the new San Juan River Plant and the junction of the new pipeline near Gallup.

At this juncture, El Paso made the decision to depart from its usual procedure of doing its own construction work. Between major jobs, the crews were idle, and even

for short periods, the overhead was extremely high.

Oklahoma Pipe Line Constructors handled 150 miles of thirty-inch pipeline from Plains, Texas to a point near Corona, New Mexico. R.H. Fulton contracted for 252 miles from Corona to a spot near Gallup. Western Pipe Constructors completed 165 miles of line to Kingman, Arizona. In addition, El Paso's remaining crews laid gathering and field transmission lines in the Spraberry gas production area of West Texas.

Of the total 400 million cubic feet of gas a day, 300 million came from the Permian Basin and 100 million from the San Juan Basin. Increased deliveries of 150 million cubic feet a day were ultimately made to Southern California Gas Company and Southern Counties Gas Company. Under this expansion, Pacific Gas and Electric Company also received an additional 150 million, and customers in West Texas, New Mexico and Arizona got another 100 million cubic feet per day.

The new pipeline, called the Permian-San Juan Crossover Line, aside from providing the additional volumes of gas needed by the distribution companies, gave the Company a new flexibility. By directly connecting the Permian and the San Juan basins through the new thirty-inch crossover line, gas reserves from all sources of supply were made easily available for delivery to all points of demand. When production of residue gas in the Permian Basin was low, El Paso was able to draw heavily on gas reserves in the San Juan. And when residue gas production in the Permian was high, it was easy to curtail withdrawals from the San Juan.

To process the casinghead gas from the Permian Basin, three new plants were constructed south of Midland, Texas; Midkiff, Pembrook and Driver plants took over the chores of gasoline absorption, fractionating, dehydrating and compressing the gas for its trip westward. Three new reciprocating compressor stations, Plains, Lincoln and Bluewater, were built to move the gas from Yoakum County, Texas, to Gallup, New Mexico.

As an added bonus to science, the crossover line provided a profile of American Pueblo Indian history and culture as hundreds of miles of ditching provided an opportunity for archeologists to unearth thousands of artifacts and dozens of prehistoric sites.

THE EXCHANGE AGREEMENT WITH NORTHERN NATURAL

Like the old cliché about "taking coals to Newcastle," a similar situation reared its head in the Texas Panhandle.

It all started when Northern Natural Gas Company embarked on a large-scale expansion. Northern Natural, with headquarters in Omaha, Nebraska, was then delivering about 825 million cubic feet of gas a day to Midwestern markets in Oklahoma, Kansas, Nebraska, Iowa, South Dakota and Minnesota.

In 1953, Northern's main source of

EL PASO'S SKYLINE UNDERGOES A STRIKING CHANGE

CORNER OF TEXAS AND STANTON streets in downtown El Paso
looked like this shortly before Borden's Ice Cream Parlor and other businesses
were leveled to make way for construction of El Paso Natural's new building.

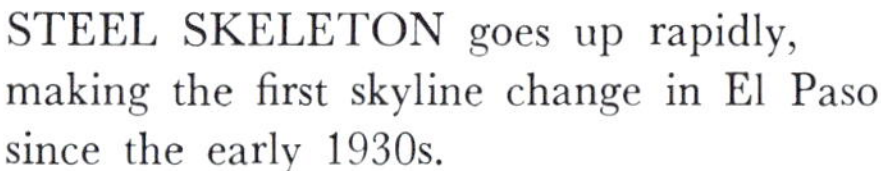

STEEL SKELETON goes up rapidly,
making the first skyline change in El Paso
since the early 1930s.

AS DEMOLITION PROGRESSED,
horse-drawn wagons from Juarez picking up
scrap were familiar sights.

198

HARD-HATTED steel workers apply
finishing touches to building's exterior.

EIGHTEEN STORY El Paso Natural
Gas Building held official grand opening
on March 4, 1955.

supply was the Texas Panhandle Field and the Hugoton Field. Under its expansion plans, Northern turned southward toward the Permian Basin. A subsidiary of Northern Natural had recently been granted permission by the FPC to construct facilities in Lea County, New Mexico and in the Spraberry area south of Midland, Texas to transport additional quantities of gas.

The next step, of course, was to plan a pipeline from the Permian Basin north to Northern Natural in the Texas Panhandle. Northern would then be able to take over the gas and pipe it to its Midwestern markets.

A startling coincidence in the whole project was the fact that El Paso Natural Gas already had a pipeline extending from Dumas, in the Panhandle, to Lea County. This line, built in 1947, was being used to develop a greater supply of gas to serve the markets of El Paso Natural in West Texas, New Mexico, northern Mexico and California.

If the new pipeline proposed by Northern Natural had been constructed in the Panhandle, the result would have been two pipelines running parallel to each other for more than 200 miles in opposite directions, one carrying gas north and the other south.

To avoid this unnecessary duplication of facilities, officials of El Paso and Northern quickly went into a huddle. They came up with a simple swap. Under the agreement, El Paso would reverse the flow of gas in its Dumas line. The Company would then pipe the gas northward instead of southward. In return, the Permian Basin Pipeline Company would deliver an equal amount into the line of El Paso Natural Gas Company at its new Plains Compressor Station.

El Paso's Dumas Station was located only fourteen miles away from Northern's Sunray Station. Under the plan, Northern built a connecting link between the stations, thus picking up gas from El Paso and piping it into its regular lines destined for the Midwest.

While all this far-reaching expansion was going on in 1953, construction was entering its final stages on several projects which were part of previous expansion programs. These projects were responsible for the building of such installations as the ten turbine stations which by now were in operation along the Southern Mainline. Many of the Company's existing plants and compressor stations in the Permian and Southern divisions were affected by this expansion as additional horsepower, cooling facilities and treating plants were installed.

In the San Juan Division, seven new plants were built to increase needed gas deliveries from the San Juan Basin. Gallup, Leupp, Williams, Kutz and Angel Peak compressor stations went into operation. Blanco Gasoline Plant and Wingate Fractionating Plant were nearing completion.

The decade of the Fifties was indeed a golden age for El Paso Natural Gas Company. Pipelines were being built at an unprecedented rate to supply energy to new industries and millions of people who were moving into California and the Southwest. The Company had become one of the nation's largest transmission firms and its success showed up boldly in its financial statements to stockholders. Its total assets jumped from something over $400 million in 1952 to almost $700 million in 1953. It seemed the only way to go was up.

But dark clouds were gathering on the horizon. Twenty years before Americans ever heard the phrase "energy crisis", Paul Kayser could foresee the first warnings of the future. In 1953, Kayser wrote to stockholders that natural gas would someday be in short supply. El Paso Natural began making studies on coal gasification for future energy supplies.

Although gas deliveries to southern Nevada (to Nevada Natural Gas Pipeline Company) began for the first time in 1954, the year had been comparatively quiet. But the lull — if it could be called that — had been short. For a company where the tensions, late hours and excitement of an almost constant program of expansion had been standard fare for years, the first few months of 1955 had seemed almost quiet.

With the beginning of the year the Company had made its complex move from offices scattered all over downtown El Paso into its new Home Office Building. An ambitious exploration program was in high gear, and a subsidiary — Rare Metals Corporation of America—was busily providing evidence of what seemed like a bright future in the form of newly built and operating facilities.

In the summer of 1955, the tempo increased. Lights burned late in the Home Office, and there were hurried comings and goings, conferences. Something was up, and when the announcement finally came, it turned out to be something big. El Paso Natural revealed that it had asked the FPC for permission to increase gas deliveries by 450 million cubic feet a day and to make the necessary expansions to do it.

Meanwhile, the Company had entered a whole new field: The refining and marketing of its natural gas by-products. The first announcements were followed by related events and developments, the sum of which made it clear that there was going to be more than enough to keep everybody occupied for quite a while. The "lull" had been no more than a chance for El Paso to catch its second wind before taking on other, bigger tasks.

THE PIPELINE ARCHEOLOGISTS

A THOUSAND YEARS AGO a great civilization burst into full bloom in the American Southwest. This was the golden age of the Pueblo Indians who lived in villages scattered throughout parts of present day New Mexico, Arizona, Utah, Colorado and the Mexican states of Chihuahua and Sonora. These hardy people lived in great multi-storied buildings and perfected a high level of architecture and ceramics.

When the Spaniards, led by an indomitable conquistador named Francisco Vasquez de Coronado, first broke into the Pueblo world in 1540 they found eighty inhabited towns. With a force of 300 Spaniards and several hundred Indians he set out to search for the Seven Cities of Cibola that were said to be stocked with gold and gems. He found no golden cities, but he led his army to the Old World's discovery of the Continental Divide, the Grand Canyon and across the Great Plains to Palo Duro Canyon near present-day Amarillo, and on into central Kansas. Today the Pueblos are reduced to some thirty villages, most of them in the valley and tributary drainage of the Rio Grande.

The greatest decline of the Pueblo civilization, however, came long before the encroachment of the white man. For example, the San Juan River Drainage Basin of Colorado, Utah, Arizona and New Mexico actually had no Pueblo Indian population at all in 1540 when the Coronado expedition entered the Southwest. But several hundred years before, the whole valley teemed with people along the tributaries of the San Juan. At that time, great centers of community life had been developed at Mesa Verde and Chaco Canyon.

Largest of the many great pueblos in Chaco Canyon was Pueblo Bonito, which contained in its heyday more than 800 rooms and thirty-two kivas, or ceremonial chambers. These rooms were terraced upward and outward to heights of four or five stories, housing an estimated 1,200 people at one time. Prior to 1887, Pueblo Bonito was the largest apartment house built anywhere in the world.

About 1250 A.D., this classic age of the Pueblos went into a tailspin from which it never recovered, causing a general breakdown within the whole cultural structure. There occurred a wholesale abandonment of this region and the populations shifted to the south and southeast.

To account for this great movement is the job of the archeologist.

Starting with a great deal of guesswork in the latter half of the 19th century, modern archeologists set out to explain the rise and fall of the classic Pueblo people from this Four Corners region.

Little by little, the pieces have been found and fitted together like parts of a giant jigsaw puzzle. Every excavation has confirmed or refuted a theory or told us something more about this prehistoric American who was building multi-story apartment houses long before Columbus made his first voyage to America.

El Paso Natural Gas Company, in the construction of its new Permian-San Juan Crossover Line, helped make another contribution to the writing of American Pueblo history since the route of the line crossed an area which was once the center of the great Pueblo civilization.

A tailor-made profile of the history and culture of these people was provided by the large pipeline ditchers. These machines dug a trench three and a half feet wide and more than six feet deep through the entire length of this rich archeological area.

Actually, the sixty-foot right of way itself provided the greatest source of information. Prior to any disturbance of the bulldozers every inch of the right of way — from Plains, Texas to Kingman, Arizona was inspected by trained archeologists for any prehistoric site or artifact it might contain.

When plans for the construction of the Permian-San Juan pipeline were made, El Paso consulted with the National Park Service and others on methods to preserve or study the archeological values of sites along the right of way.

For the protection of ruins on public lands, Congress passed an act in 1906 which restricted excavation on government lands to reputable, scientific-educational institutions and their qualified field leaders. This is known as the Act for Preservation of American Antiquities and it sets forth procedures under which archeological sites can be examined and excavated.

To conform with the provisions of this act and carry out its purposes and objectives consistently, El Paso's C.L. Perkins collaborated with Dr. Jesse L. Nusbaum, who at that time was Senior Archeologist of the National Park Service and consulting archeologist for the Department of the Interior in all matters pertaining to the Federal Antiquities Act. Dr. Nusbaum was largely responsible for the success of previous archeological studies along El Paso's original San Juan pipeline when it was constructed in 1950 and 1951.

Dr. Nusbaum is one of the country's foremost archeologists and was a member of the expedition to Mesa Verde National Park in 1907. In cooperation with El Paso Natural, Nusbaum organized, supervised and worked closely with five archeologists who were on the Company's payroll. These men first made a surface survey to locate sites. Then they directed the excavation crews along the pipeline,

DR. JESSE NUSBAUM (left) and assistant Franklin Fenenga
inspect artifacts in Pueblo Ruins.

working speedily to keep ahead of the bulldozers clearing the right of way.

Two crews of archeologists worked along the line, supervising the actual salvage operations, making comprehensive maps of each site and inspecting each artifact found. The excavation was usually done by six men and each team was hired especially for the job. While working on the Acoma Indian Reservation in northwestern New Mexico, one team consisted entirely of Acoma Indians.

El Paso Natural paid all salaries and expenses of the pipeline archeologists, the wages of helpers who assisted in excavating sites and whatever other expenses were involved. The crews were provided Jeeps for their travel along the right of way.

The first step in the archeological study of the area was a surface survey of the entire right of way. Every mile was checked and rechecked — on foot — ahead of the advancing bulldozer crews. And every possible site was charted.

The likelihood of an archeological site beneath the surface of the ground can be detected in a number of ways. When archeologists find an area containing broken pottery, flint chips, bits of charcoal and occasional arrow points, the chances are good that a structure or camp site may exist under the surface. Another good clue is a slight rise or a mound of earth in a spot where the surrounding territory is generally flat.

Usually these sites are located on higher ground, which at one time furnished a measure of protection from predatory tribes. Sites are also usually found in greater concentration close to a good water source — especially a river valley where crops would have been raised. Many of these pueblos had been completely buried by the effects of erosion of hundreds of years of blowing sand. Occasionally, however, the archeologists found the tops of adobe or masonry walls protruding above ground.

After completing their surface survey and charting and mapping of all sites, the archeologists moved in and began excavating. At each location, artifacts were carefully labeled, measured maps were made of all features and extensive notes were taken. These were all used later in writing and publishing a 400-page book (dedicated, incidentally, to El Paso Natural Gas Company) describing the excavations and findings along the Permian-San Juan pipeline.

Following closely in the footsteps of the archeological crews came the pipeliners and their bulldozers to clear the right of way. Then came the ditching machines. Occasionally these machines unearthed and revealed in trench wall profiles a site containing artifacts and potsherds that were not discovered on the surface survey. When this happened, the archeologists moved back into the area and examined the soil dug up by the ditchers. They also inspected the walls of the trench, where it was easy to distinguish the profile of a prehistoric camp site.

When a major site was discovered in this way, the pipeline was rerouted around the area so that Dr. Nusbaum and his crews could take the necessary time to dig and excavate further.

The Laboratory of Anthropology of the Museum of New Mexico in Santa Fe received all material the archeologists uncovered in New Mexico. Anything found in northern Arizona was deposited in the Museum of Northern Arizona at Flagstaff. In the laboratories of both these museums the material was carefully catalogued and preserved for future research and educational use.

Each of the excavated sites helped complete the picture of the great parade of prehistoric occupation as the new pipeline snaked its way through areas in New Mexico and Arizona which had never before been excavated, and many new discoveries were made every day during construction of the line. These findings along the course of the pipeline proved a great boon to scientists in telling more and more about the pageant of Pueblo civilization, the first evidences of which occurred in the period ranging from the beginning of the Christian era to about 500 A.D.

This was a pre-pottery era, the age of the Basket Makers. They were experts in making artistically twined woven bags and coiled baskets, sandals, fur cloth robes, and nets and snares for catching small animals. In these early levels, archeologists found no signs of bows and arrows. There was no evidence of true fired pottery until about the end of the fifth century A.D.

With the advent of the sixth century, the bow and arrow made its appearance, thin pottery vessels were made and the people began to cluster their homes together to form small villages of semi-subterranean, earth covered pit dwellings. This period is known as the Modified Basket Maker, and it extended from about 500 to 700 A.D.

The next chapter in the Pueblo culture is called the Developmental Pueblo and dates from 700 to 900 A.D. This era saw the great development of the growing of cotton, the weaving of cotton cloth and the exclusive use of the bow and arrow. Houses were built above ground in single story units of six to fourteen rooms. They were made in varying combinations of poles, stones or adobe, prior to development of rock masonry. The Indians in this period also became adept in making many types of painted and corrugated fire-baked pottery vessels.

The Great Pueblo period was ushered in about 900 A.D. and it continued to advance until about 1250. This is the best known epoch of these people and is characterized by the classic development in masonry architecture and the making of pottery.

The highest examples of building construction were achieved in this period with the creation of terraced apartment-like dwellings. These contained as many as 800 rooms in open valley and mesa locations as in Chaco Canyon, and as many as 200 rooms in the cliff dwelling caves of Mesa Verde. There are a number of theories concerning the development of these amazing structures. Some believe that fear of raiding by enemy Indians of other tribes forced the peaceful Pueblo

COMPANY ARCHEOLOGISTS (above) unearth part of Pueblo Indian
village which was abandoned about 900 years ago. Below, Dr. Nusbaum
inspects a painted pottery bowl during digging on original San Juan line.

DR. NUSBAUM and one
of his assistants catalogue
material on the Acoma
Indian Reservation in
Northwestern New Mexico.

FRANKLIN FENENGA
makes field notes during
excavations along the Permian-
San Juan crossover line.

people to gather in large groups and to construct defensible communal houses in the open or in naturally defensible caves. Others believe that these buildings were the natural evolution and the product of an architectural vogue rather than fear of outside marauding groups.

By the end of the Great Pueblo period, various forces caused a general breakdown within the large centers of population. In archeology, this is known as the Regressive Pueblo period and it extended from about 1250 to 1600. During this time the large Pueblo structures in many areas were completely abandoned, and the population shifted principally into the Rio Grande Valley and its tributaries. This is where the first Spanish explorers encountered them, and these are the same general regions where modern day Pueblos now make their homes.

The reason for the decline of the Pueblo's golden age is still anybody's guess. The most valid explanation seems to be that recurring droughts forced the farming people to look for greener pastures. On the basis of tree ring studies and the resulting tree ring calendar dating of ruin beams, it is known that the Great Twenty-Five-Year Drought occurring between 1276 and 1299 forced the abandonment of many regions of the Colorado Plateau in the early years of that incredible dry spell. The approach of enemy people—the Western Apaches, the Navajos and the Utes—was no doubt a contributing factor, but the elemental necessity for an adequate water supply was the final and compelling reason for the exodus.

By 1540, the Pueblos seemed to be moving toward a renaissance, but the coming of the Spanish conquistadors and the subsequent conquest by Europeans ended further development of the original civilization.

According to Stewart Peckham, director of the Division of Anthropology of the Museum of New Mexico, El Paso Natural's efforts to preserve archeological sites set a pattern that the whole country was to follow in later years. "Work on El Paso's pipelines," says Peckham, "stimulated other people's interest in preserving archeological remains that were going to be destroyed by construction. And since that time archeological work has been extended to all sorts of other types of earth-disturbing activities. Whether it's power line construction, or dams or reservoirs, housing developments, coal mining—you name it—there will probably be archeological work involved. Even the work that's being done in Alaska now seems to have its foundations back there with the work El Paso supported here in the Southwest."

HIGH FINANCE AND THE FPC

How does a federally-regulated company like El Paso put together multimillion dollar projects? Answer: Very carefully—and often in spite of frustrating delays that may go on for years. Nonetheless, it's all part of the system under which El Paso (as well as all other interstate pipeline companies and electric companies) works. But the utility concept of providing services for the American public is recognized as one of the most successful practices of the U.S. economy largely because it avoids the expensive duplication of facilities.

If an American city were served by five electric companies, say, the result would be a great tangle of overhead wires, expensive and duplicated offices and warehouses and expanded operating expenses. The Federal Power Commission certificates an organization to provide energy to a certain area and then closely regulates the company in carrying out its job. El Paso Natural, which transports and sells gas in interstate commerce, is regulated by the FPC. There are five FPC commissioners, appointed by the President of the United States. The Commission is bipartisan and generally made up of three members of the majority party in Congress and two of the minority party. The FPC staff is much larger, having about 1,000 employees.

The Natural Gas Act of 1938 gave the Commission jurisdiction over the interstate transportation and sale of natural gas. In 1954, the U.S. Supreme Court ruled that the rates and sales of independent producers selling natural gas in interstate commerce are subject to FPC jurisdiction.

Travis Petty, who became President of El Paso Natural in 1976 (and who began his career with the Company in 1946 in the mail room), sums up the evolution of a typical project this way: "We start out with the germ of an idea which involves a new pipeline or expansion of facilities and additional sales. Back in the earlier days Mr. Kayser would usually match a new gas supply with a market. Ordinarily, these things are kicked off by an unusual demand in the market place that provides an impetus to go out and make some extraordinary effort to get new gas supplies. Or, conversely, we convince somebody that we have sufficient supply and that we have a market on the other end that can take the gas on economically attractive terms. The supply and the market are the essential ingredients, and we have to satisfy ourselves that we have both.

"After the preliminary agreements are worked out with a customer, we have to demonstrate through appropriate studies and exhibits to the FPC that the proposal is feasible and in the public interest. At the same time we are giving consideration to where the money will come from. That's really where the project begins to be put together.

"The proposal is then filed with the FPC and it is made available to the general public so that everybody who may have an interest is afforded an opportunity to be heard in formal hearings. The Commission is generally very flexible, very liberal, in its attitude toward people coming in and participating in these hearings. That doesn't mean that a fellow walks in off the street and decides he wants to participate in a hearing. But generally, a state commission, any public body, a customer or a supplier who may be affected economically by what we propose is usually allowed to get in and participate.

"Once you have all the pieces, the basic ingredients—the supply contracts and the market contracts, the project is pretty well defined and you know exactly where you're going. Back in the Fifties, during the period of most rapid expansion, we were filing these things one on top of another. We could crank out a pretty hefty application in a period of six to eight weeks. That involved a great amount of work by many people, particularly the Engineering Department that had to be responsible for maps, flow diagrams, cost estimates and dozens of other things."

All of this material is printed and bound into individual copies, and in years past, fifty or so copies were all that were needed for an application. "But," says Petty, "our life has become so much more complicated in recent years and we operate in such a broad area that there are an infinite number of people who show an interest in our filings. More recently we printed something like 1,500 huge volumes (each about a foot thick) for distribution. Our Printing Department never ceases to amaze me with how much work they turn out. They worked on one application recently, literally around the clock for about six or eight weeks, just physically printing the material that we were feeding them."

The time lag from first filing to a final decision is not nearly so fast as preparing an application and it is considerably more frustrating. On a particularly controversial matter these decisions have taken from about one year to a maximum of nine and a half years, with the median time being about three years. "Certainly," says Petty, "the FPC is not necessarily at fault with these situations, although we have been terribly frustrated from time to time with the delays in getting answers. Because of the FPC's requirements to admit people with legitimate interests, the administrative proceedings and then possible court appeals, there are just certain handicaps they can't overcome. They have to operate under the law, and under the best of circumstances the system is slow and

W. Burney Warren

cumbersome. You complicate that with a large number of intervenors—a lot of them hostile and deliberately trying to delay one of our projects. So the bigger the project, the more controversial, the more complicated it gets, and the time requirements grow geometrically."

The financing of these projects, especially when you need hundreds of millions of dollars, is a story in itself. It comes down to one word—money. Burney Warren, The El Paso Company's Executive Vice President in charge of finance and Treasurer of the Company, has played a major role in the money market for many years. According to Warren, "During the period of El Paso's rapid growth we had all the ingredients that were needed to attract investors, but it's getting tougher all the time. You have to have a market. We have always had that and still have. You need an organization that the investing public has confidence in, and of course, you have to have a gas supply.

"It's interesting to look back and see how far we've come. When we started our first project to California we were a $30 million company. The California line was estimated to cost in the range of $42 million, and before we really got started, we piled a second phase on it and we were up to around $68 or $70 million. So here we were starting on a project that more than doubled the Company financially. At that point there was no way we could finance the pipelines except on a project basis, which meant you arranged the financing for it, and through the mechanics of negotiating (which was done through insurance companies) you got your money—not all at one time, but over several closings with the insurance companies. Money from the insurance companies along with the required amounts of equity furnished by El Paso went to a trustee—like Manufacturers Hanover Trust Company. The equity required for these projects was raised through the public sale of common and preferred stocks and earnings retained in the business. The trustee then paid for expenditures as we made them. We would deliver to them certified statements with an engineer's report that the work had been done and we had spent a certain amount on a part of it. Then we'd get the money and start over again.

"That type of project financing went on through the Fifties primarily because we were stacking one project on top of another to where we never had to slow down. Each one was bigger than the last one. Then about 1960, it was felt that the Company had matured so that we went on what is a far more normal type of financing. We handled our construction with short term bank loans, and supplying our equity (the amount in our capitalization that is represented by the ownership of stockholders) as we went along. Then when we had done a certain amount of construction and our bank credit lines were pretty well used up, we would do a permanent financing, pay off the banks and start over again.

"We have our bank line of credit in six banks—Chase Manhattan, First National City Bank, Manufacturers' Hanover, Morgan Guaranty, Mellon Bank and Continental Illinois Bank.

"The insurance companies that hold the major part of our debt are Metropolitan, Equitable Life, New York Life, John Hancock, Travelers', Aetna, Mutual of New York and Northwestern Mutual. The significant thing about this is that the relationship has been built over a period of many years, going back to Mr. Kayser's dealings with them. At one time, Mr. Kayser did all the financing and the relationship that he built up with these companies is something you can't put a dollar value to, but it has tremendous value to the Company."

EL PASO
ELPA
EL PASO
EL PASO

10

BORN RUNNING

El Paso Natural Gas

Products Company

BACK IN 1907, a German by the name of Herman Blau cooked off a batch of crude oil, condensed its vapors and came up with a rather strange liquid. Herr Blau discovered that gas from this strange, evil-smelling product burned with a bright hot flame.

Overjoyed, he could picture hausfraus all over the world using it as fuel in their kitchens, and the stuff was hurriedly placed on the market. Sadly enough, it proved to be generally unsatisfactory as a fuel, and the whole venture was junked after enjoying only limited success. Its equipment was too complicated, and besides it cost too much. The production of "liquid gas" still had a long way to go, but it did prove one thing — Herman Blau was on the right track.

Ten years and dozens of trials and errors later, the production of liquefied petroleum gases was fast coming into its own, to become a major industry. By 1955, El Paso Natural Gas Company was producing a million gallons of liquid products a day from its natural gas stream in the form of butane, propane and natural gasoline. Never-ending orders for butane and propane poured into El Paso's Home Office, and tank cars of butane and propane were shipped to widely separated places like Sioux City, Iowa; Tampa, Florida; Cadillac, Michigan; Ontario, Canada and Mexico City.

Four years earlier, when the San Juan Plant was completed, natural gas production began to climb skyward. This meant a large increase in the amount of liquids that had to be separated from the gas to make it saleable. The liquid surplus skyrocketed again with the completion of Jal Plant No. 4 in 1952. Liquid petroleum gas is a combination of hydrocarbons known as propane and butane. The properties of these liquids are rather unusual; they can be stored and transported in liquid form, but when released from storage they vaporize and can be burned and used as a gas.

At El Paso's natural gas purification plants, propane, butane and natural gasoline are extracted from the natural gas as it comes in from the field gathering lines. Most of the gas the Company produces and processes contains these liquids vaporized within it. On an average, each million cubic feet of natural gas will contain about 1,000 to 4,000 gallons of liquid hydrocarbons. Left in the natural gas they tend to clog the pipelines and reduce transmission efficiency of the gas to the consumer. Extracted from the natural gas, however, the liquids perform some very valuable functions. Although not quite as economical as natural gas, butane and propane have hundreds of uses. They serve farms and homes which are too far from natural gas pipelines to enjoy natural gas service. They can be hauled and stored as liquid under pressure, but are readily converted to gas and move through pipelines and burners as gas when pressure is reduced. In some cases, butane and propane serve as fuel for large industrial engines as well as for large trucks, buses and tractors.

Butane also figures prominently in the huge appetite of the synthetic rubber industry. From butane comes the petrochemical butadiene, one of the industry's basic raw materials. The petrochemical industry puts natural gas liquids to some of their most interesting uses, particularly in the field of plastics. Polyethylene squeeze bottles, for example, can trace their ancestry directly to propane.

In making polyethylene, chemical plants treat liquid propane and break it down into a gaseous substance called ethylene. The ethylene is then treated by extremely high pressures which completely change its molecular structure, and it becomes a sticky, doughy translucent mixture. This is raw polyethylene. It is later extruded into a long, flat ribbon, and as it dries, it becomes rather hard. Then it is chopped into pellets about an eighth of an inch square and is sold to the fabricator in this form. The manufacturer melts the pellets and molds them into any shape he wants—bottles, garden hose, or any one of hundreds of different products.

Although propane and butane together represent the greatest percent of natural gas liquids El Paso produces, another valuable product derived from the gas stream is natural gasoline. Its chief use is in the blending of motor fuels, and by 1955 the refiners of gasoline were purchasing almost eighty-seven million gallons of natural gasoline from the Company. Natural gasoline normally has a higher vapor pressure (the ability to change very quickly from liquid to a gaseous state) than other motor fuels. During winter months this is a great advantage in starting an automobile engine on a cold morning.

From its gasoline extraction plants in the

Permian and San Juan basins, the Company was producing some 25,000 barrels a day of butane, propane and natural gasoline, and the volume was increasing every year as new gas, by the millions of cubic feet was being processed and fed into the mainlines for western customers. Since these by-products had a rather weak and seasonal demand, the Company was beginning to see the dismal prospects of swimming in a great sea of liquids.

The answer was the formation of the Company's newest offspring, El Paso Natural Gas Products Company. "This company wasn't simply born, it was born running." Thus did an official of the newly organized subsidiary explain the rather hectic pace of the fledgling organization.

Most new firms, after life is breathed into them, can be coaxed and coddled along until they can stand on their own two feet. Not so with El Paso Natural Gas Products Company. Even before the firm had printed stationery, it was doing business on a lively scale. The reason—the new company immediately acquired and assumed operation of three crude oil refineries—the McNutt Oil and Refining Company near El Paso (in October, 1955) and two smaller units purchased from Malco Refineries at Prewett and Bloomfield in the San Juan Basin. The McNutt refinery, built in 1933, had a capacity of 3,000 barrels of crude oil a day and it controlled 125 retail marketing outlets and bulk stations in West Texas, New Mexico and Arizona.

Paul Kayser was named President of the new company; C. L. Moore, Vice President and general manager; and Cecil McNutt, Vice President and operations manager. Employees of the Products Company started moving into their quarters on the fifth floor of El Paso's Luther Building in February, 1956. Most of the personnel needed, in addition to employees of the acquired companies, were drawn from the parent firm El Paso Natural Gas.

At the time, all of the gasoline was marketed under the "Dixie" trade name, acquired from McNutt. Dixie "regular" sold for twenty-six cents a gallon, about three cents a gallon less than most of the major oil company stations.

Several years before, two key personalities who were to have a tremendous impact on the Products Company appeared on center stage in Odessa, Texas, the capital of the Permian Basin energy empire. They were W. D. Noel and Jesse Owens. Owens was a chemical engineer who also happened to be a manager of the Odessa Chamber of Commerce. Together with a group of civic and business leaders headed by Noel (an independent oil and gas producer and a millionaire by age twenty-seven) they systematically visited dozens of industrial leaders all over the country. In two and a half years they were to travel almost 100,000 miles in their efforts to sell two things: Availability of cheap raw materials and a productive work force. The idea was to use the abundant by-products of the Permian Basin to produce butadiene, the basic ingredient of synthetic rubber for tires. Oddly enough, the man they were looking for, in

W. D. Noel

a manner of speaking, was right in their own backyard.

The right man was Paul Kayser, founder and then Chairman of the Board of El Paso Natural Gas Company. In 1954, Noel approached Kayser with his proposal. According to Noel, Mr. Kayser said, "Bill, I don't even know what butadiene is!" "But," says Noel, "I went into some explanation and he thought it sounded all right. His approach to it was very similar to the approach he had used in building El Paso Natural. Kayser said, 'Bill, if you can get us a long-term take-or-pay contract from a major rubber company, El Paso would be interested in joining you and your associates in making this investment.' "

Because abundant water supplies are needed for plants like this, Owens and Noel had some difficulty explaining to major rubber companies that such a plant could be built in landlocked Odessa, far from most petrochemical plants which are located near areas such as the Gulf Coast of Texas and Louisiana.

"Jesse and I finally got General Tire and Rubber Company very interested in building a rubber plant in Odessa," says Noel. "These negotiations continued until it was time to take it to Paul Kayser and the El Paso people. But in the meantime, a competitor showed up on the horizon and was also talking to General Tire about building a butadiene and styrene plant.

"It wasn't all cut and dried for us," Noel emphasizes, "but we finally had the contract commitments orally developed to the point that we were ready to sign the agreements that would create the beginning of a petrochemical complex.

"We were negotiating primarily with Mr. Bill O'Neil, the Chairman and founder (now deceased) of General Tire. One General Tire executive was leaning a little bit more toward our competitor. But one cold afternoon in Ohio, we were in Mr. O'Neil's office; Mr. Kayser was there, Howard Boyd was there, we were all there —and the contracts and oral agreements had been drafted and were ready for execution.

"The General Tire executive said: 'Bill, before we sign those contracts I think we ought to look at this other proposal.'

"Here's where Mr. Kayser made the most beautiful move I've ever seen. He looked up, moved toward the edge of his chair, looked around and said, 'If we are going to talk about it, where's my hat?'

"Bill O'Neil said, 'No, no, let's go on with the business.' But we were that close to having the whole thing blown apart. It was 'where's my hat?' that saved it. The butadiene contract, the styrene contract and the agreement to build the rubber plant in Odessa were signed that day by El Paso and General Tire."

By June, 1956, bulldozers sent up dense swirls of dust as they rooted out stubborn mesquite and greasewood on the outskirts of Odessa. Nearby a small group of townspeople watched the operation with interest, for it marked the entrance into their city of a new multimillion dollar petrochemical business — that was soon to become the largest inland petrochemical complex in the

DIXIE service station sign marks the location of Products Company offices in El Paso's Luther Building.

United States. The butadiene and styrene plants, along with General's rubber plant, were built and went into production in 1957 and 1958.

During those years Gas Company employees were offered jobs with the fledgling Products Company, and El Paso's Luther Building was filling up with accountants, engineers, draftsmen, secretaries and a number of new job classifications like marketing and sales promotion. It was a whole new realm for most of them, since they were now dealing with the general public for the first time. The forty-hour week became a fond memory. They learned quickly not to say "filling station" any more. Instead, the right name was "service station." They learned about gasoline price wars, bulk plants, credit cards, and that the best location for a new service station was on the "goin' home" side of the street.

At first, all of the retail outlets were Dixie Stations competing with other independent gasoline marketers, but with the building of two sparkling new refineries—one at Odessa and the other at Ciniza (near Gallup, New Mexico)—the Products Company introduced a new brand of gasoline, El Paso Red Flame.

Pennants flew, bands played and pretty girls distributed gifts. A festive holiday spirit prevailed in Farmington, New Mexico, caused by the 1957 coming out party in mid-December of El Paso gasoline.

A name well known in Farmington since El Paso Natural had set up headquarters for its expanding San Juan Division some six years before, El Paso seemed a fitting

PRODUCTS COMPANY butadiene plant as it looked in 1958, shortly after completion.

ADVERTISING PROMOTED RETAIL PRODUCTS THROUGH 1,450 SERVICE STATIONS. THE MARKETING AREA EXTENDED FROM SAN ANGELO, TEXAS ON THE SOUTH TO YUMA, ARIZONA ON THE CALIFORNIA BORDER, AND FROM THE MEXICAN BORDER TO SOUTHERN COLORADO AND UTAH.

TOMORROW'S GASOLINE TODAY

EL PASO

As compression ratios rise, more and more late model cars are bothered with knock, ping and rumble. Now EL PASO engineers bring you the perfect solution — EL PASO Red Flame *SUPER* Premium Gasoline, containing Isotane, the alkylate component of powerful aviation fuels.

RED FLAME
SUPER PREMIUM GASOLINE offers you flashing new performance, brilliant new driving pleasure! You actually can FEEL THE DIFFERENCE in the way your car runs!

PRETTY GIRLS helped promote El Paso
Red Flame motor oil. Above, Mel Blanc, better know
as the voice of Bugs Bunny, appeared in television
and radio commercials for the Products Company.

VARIETY of plastic and rubber products made from materials produced at Odessa Petrochemical Complex.

GENERAL TIRES were promoted and sold through Products Company's service stations.

EL PASO RED FLAME gasoline made its debut in December, 1957 in Farmington, New Mexico.

title for the Company's newest infant. So the new gasoline made its debut; a big red flame and the words El Paso appeared on interior lighted plastic signs in front of two spanking new service stations.

During the open house held December 14 and 15, visiting Farmingtonians were presented with mementos such as orchids, perfume, key chains, sleet scrapers and comic books and they registered for drawings on TV sets, diamond wrist watches, hunting rifles and bicycles to be presented with the compliments of El Paso gasoline.

New stations were built to accommodate the new brand of high octane gasoline which was as good or better than anything the majors had to offer. For some time, the retail outlets had a double image: they were called El Paso-Dixie service stations.

In addition to the Dixie outlets, the Products Company bought the International Petroleum Company, an El Paso, Texas firm with nineteen retail outlets in El Paso. Other phases of the Company's business in El Paso included the El Paso General Tire Service, a large General Tire store specializing in brake and wheel adjustments, and the El Paso General Appliance Center, where a complete line of name brand products were sold.

The rapid growth of the Products Company and its expanding market area provided ample testimony of the company's boldness in departing from the usual, tradition-tied way of doing things. When the company was organized in 1955, it had fewer than 100 employees. By 1957, the

STYRENE PLANT at the new Odessa Petrochemical Complex in 1958.

firm's payroll listed 1,000 employees, 250 retail outlets and fistfuls of new ideas and projects. C. L. Perkins became President of the Products Company.

In 1958, Donald B. Thurman, who had been in charge of marketing in Latin America for Standard Oil of New Jersey, took over the reins as marketing director of the Products Company's petroleum outlets. "At that time," says Thurman, "we were selling 7,000 barrels of petroleum a day; about half of which was gasoline and the rest diesel fuel, heavy fuel oil and sand frac oil. It was kind of a tricky marketing operation because the Company had Dixie Stations, El Paso-Dixie Stations and the new El Paso Stations. The El Paso Stations were in direct competition with the majors, although they tried to stay about a cent under the price of the majors. The two new refineries at Odessa and Ciniza each were refining some 8,500 barrels of crude a day and the same quality gasoline was distributed to all of the Company's retail outlets. In some instances, there would be an old Dixie Station on one corner and a bright new El Paso Station across the street."

Thurman chuckles as he recalls the time he drove into a Dixie Station and asked the driveway attendant what the difference

228

GRAND OPENING of the Odessa Butadiene Plant was
attended by El Paso Natural officials; among them were
Mr. and Mrs. Virgil Rittmann at left, and Mr. and Mrs.
R. W. Harris at right.

CATHERINE MARTCH and
Lambert Moore were on hand for
festivities at Odessa Complex
opening in 1957.

was between gasoline sold by Dixie Stations
and El Paso Stations. "About three cents,"
answered the attendant.

As the 1950s came to a close, the Products Company continued on its road to
rapid expansion, both in production and
refining of crude oil and manufacturing
petrochemical products. In 1959, sales of
refined products from the Odessa and
Ciniza refineries were above expectations.
The capacity of the Odessa refinery was
increased from 9,000 to 13,000 barrels a
day.

New to the Southwest was a method of
gasoline marketing introduced to its service
stations. Called "Precision Blending," it
involved marketing six different grades of
gasoline from a single, highly accurate
blending pump. The blending pumps were
fed from two storage tanks, one containing
a gasoline called El Paso 400 and the other
containing the Red Flame Super Premium
480 gasoline. A selector dial on the pump
permitted the attendant to change the proportions of the two fuels so that he could
deliver any of six different gasoline blends
with varying octane ranges, each at a different price. A seventh grade of gasoline,
Dixie 390, a regular grade gasoline was dis-
pensed from a separate pump. (One com-

BUTADIENE PLANT at the Odessa Petrochemical Complex.

pany advertising official turned purple when he was greeted by an attendant who said, "What flavor do you want, fella', chocolate or vanilla?)

Also in 1959, the Products Company expanded its string of service stations by contracting or acquiring 110 stations, and by the end of the year it had 455 outlets selling its brands of gasoline and motor oil in an area extending from San Angelo, Texas on the east, to Yuma, Arizona on the Arizona-California border, and from the Mexican border on the south into southern Colorado and Utah.

At the petrochemical complex in Odessa, the butadiene plant completed its second full year of operation in 1959, producing butadiene for the manufacture of synthetic rubber. The plant was operating at its full capacity of 50,000 tons a year. The styrene plant also operated at full capacity, producing 40 million pounds of styrene during the year.

In an effort to provide more and more of its own crude supply for the refineries, the Products Company expanded its exploration and drilling operations far afield— from the Green River Basin in Wyoming to Lake Maracaibo in Venezuela.

For a company only four years old, success seemed just around the corner. But as the Fifties gave way to the Sixties, there were rough times just ahead. The next decade was to be a time to try men's souls, particularly those Products Company people who had been working so hard to build a vital, prosperous subsidiary for El Paso.

DIVERSIFICATION

AFTER THE GOLDEN AGE of the Fifties, which saw El Paso's new natural gas pipeline projects literally piled up one on top of another, the Company began to go through a long dry spell with the Federal Power Commission where things were becoming tighter all the time. El Paso realized the fact that having all its eggs in this regulatory basket could create future economic problems.

According to Lloyd Varenkamp, Treasurer of El Paso Natural, "We thought it might be well to diversify into certain nonregulated type opportunities. Although we didn't have anything too clear cut at the time, it was obvious that we should look at businesses that were closely related to our own. We had a small group: George Carameros, Bob Montgomery and myself. Just a little nucleus of a group looking into opportunities to diversify our operations. At that time we were really looking and were interested in almost anything, as long as we felt it was something that we could acquire and not overly dilute our management efforts or our financial situation.

"As we see diversification now, I would say it started with El Paso Products Company. The petrochemical business was a very natural kind of vertical diversification. As an extension of petrochemicals, we were interested in further upgrading the basic petroleum products, and in the mid-Sixties the Products Company went into a joint venture with the Beaunit Corporation, a major manufacturer of synthetic fibers and textiles. Ultimately, Beaunit became a wholly-owned subsidiary of El Paso Natural in 1967.

"It was Mr. Kayser's desire — with the foresight he has — to look at certain other energy sources, since he felt that someday natural gas would be in short supply. We got into Rare Metals as a means to assure an alternate source of fuel. This ultimately led to our Mining Division and later, to the acquisition of the Lakeshore Copper Mine and Narragansett Wire Co. It was in 1967 that we acquired Narragansett, and for the amount we originally invested in it, it has had some very good years.

"In 1971, we entered into land development — a joint venture for the development of a 320-acre residential community near Phoenix known as "The

DIVERSIFICATION started with El Paso Products Company

Lakes," and later jointly developed "Lake Country," an adjacent area.

"Another venture that we entered is what we call Resources Conservation Company, that we are jointly into with the Boeing Company and Reading and Bates. This operation is concerned with reclamation of brackish water. It is something that could make us some money, some substantial money a little later on, we just don't know.

"Even further removed from the energy business was our entry into the insurance business. Operating under the holding company name of American Corporate Resources, Inc., are two principal operating companies — Desert American Agency, Inc., a property and casualty general agency corporation, and Desert Insurance Company, Ltd., a foreign based reinsurance company — and they continue to be a profitable diversification."

11

RARE METALS ADVENTURE

THE COLORADO PLATEAU, in midsummer 1955, was teeming with activity. Here centered America's frantic search for uranium—a boom which had been bringing in hundreds of new faces every month since the big uranium rush got under way in 1952.

By comparison, said the experts, the Gold Rush of '49 was an Easter egg hunt. 1955's version, a mass migration to one of the last frontiers in America, brought with it not only thousands of uranium prospectors, but engineers, highway builders, truckers, outfitters and various other businessmen as well.

Bright new items of equipment, like Geiger counters and scintillators, appeared in the fly-specked show windows of cross-roads general stores. Trailer villages blossomed overnight in some of the West's most isolated back country.

The fast and furious claim staking of the past several years had also brought a boom to the stock brokerage business. In Grand Junction, Colorado you could buy a uranium claim for a few dollars, shares of stock in a uranium company for a penny a share. There was no guarantee, of course, that you'd become rich overnight.

The Colorado Plateau covers about 107,000 square miles of vast tableland that spills out into four states: Colorado, Arizona, Utah and New Mexico. By the summer of 1955, canyons and mesas were swarming with prospectors. They came in Jeeps, pickups and hot rods from all over the country. Some were college boys out on a lark, some were white collar men and their families combining their vacations wtih two weeks of prospecting. Others were drifters always ready to pick up a fast buck. Most were armed with Geiger counters, maps, sleeping bags, snake bite kits and other equipment necessary for 20th century prospecting.

The Cold War situation with Russia had reached a critical stage and the U.S. Government offered ample incentives to individuals as well as major companies to build up a stockpile of uranium — just in case the Cold War warmed up.

Against this frenzied background, El Paso Natural Gas Company's newest subsidiary was organized: Rare Metals Corporation of America.

Rare Metals had its start in 1954 when Paul Kayser called on the U.S. Bureau of Mines in Washington, D.C. His mission was to explore the means of organizing a uranium company. With El Paso Natural's geologists and exploration crews roaming the West, and thousands of acres of land under geologic evaluation for oil and gas, his interest in uranium was a natural outgrowth of events. The Company's pipeline trenches cutting across country, its seismic crews at work over thousands of square miles of desert and mountains, the drilling of its wells — all gave a close look at a lot of country, at depths of from five feet to thousands. With comparatively little additional equipment, tests could be run easily to detect the presence of uranium, along with the regular work. Chief of the Rare

Metals and Precious Metals Branch of the Bureau of Mines and supervising the agency's investigation of radioactive mineral deposits at the time of Kayser's visit was Mitchell H. Kline, a top mining engineer and uranium expert.

In 1954, El Paso's Board of Directors met in New York. The result of the meeting was the formation and incorporation of Rare Metals Corporation of America on May 24, 1954. The new company was organized by El Paso Natural and Western Natural Gas Company. Mitchell Kline resigned his position with the Bureau of Mines to come to work for the newly-formed corporation.

Richard L. McConn (who was later to become a Vice President of El Paso Natural) was transferred from his position in El Paso Natural's accounting department to Rare Metals early in 1955. "At the time," recalls McConn, "we were running Rare Metals out of El Paso, and at first it was nothing more than a set of books. That was right up my alley, but frankly I had never even been in a mine before. But the company got started. Kline brought with him some other Bureau of Mines people, like Ed Carlson, who was superintendent of the Bureau's exploration for rare metals; John Reynolds, a top mining engineer and Allen McKinney, a metallurgical expert. Then a little later we got some other men who had worked for the Bureau of Mines but also had some experience with independent mining companies — Claude Barron and Harold Horst. We also tapped Ruben Kronstadt, who later ran our laboratory and did

236

all the sampling for us.

"In September, 1955, we moved everything to Salt Lake City. I took Earl Peck and Windsor Nordin from the Gas Company and the three of us went up there and opened up an office. We hired payroll people, purchasing people, a whole office staff. Kline hired a land man, Bob Baldwin, who was already working for El Paso Natural in Salt Lake at the time we were getting set up.

"As far as officers were concerned, the Treasurer of Rare Metals was Virgil Rittmann. The Assistant Secretary and Treasurer — there were two of them — were Burney Warren and Catherine Martch. Paul Kayser was President and Howard Boyd, C. L. Perkins, Fred Wagner, H. F. Steen, D. H. Tucker, R. J. Crowley and Mitchell Kline were Vice Presidents. Crowley was previously Assistant Secretary and Assistant Treasurer of El Paso Natural until he moved to Salt Lake City where he played a major role in the development of Rare Metals.

"I went to Rare Metals with the title of chief accountant. It wasn't very long before it became obvious that that was not the right title, because I was getting involved in all administrative matters, and Windsor Nordin became, in effect, the chief accountant. I functioned more like an administrative assistant to Kline, helping him in negotiations and economic studies. Mitch Kline was really the chief honcho in Salt Lake City."

After things got put together, the first big move of Rare Metals was the acquisition of all of the capital stock of an outfit called Arrowhead Uranium Company and Arrowhead's properties near Cameron, Arizona. Arrowhead, a small "pick and shovel" operation, had been producing large amounts of uranium ore from part of its operations on the Navajo Reservation for two years. In addition, Rare Metals requested and received an exploration permit on 28,000 acres of adjoining land on the Reservation.

Tribal codes in regard to exploration on Indian lands were strictly adhered to by Rare Metals, and the coopeation the company enjoyed with the Navajo Tribal Council was partly due to the good will generated in past years by El Paso Natural Gas Company in its dealings on the San Juan pipeline.

Exploration on the Reservation was subject to strict limitations: For example, from the date a permit was issued, a company had only 120 days in which to search for and define the exact limits of the claims it wanted to mine.

This meant an ultra-high speed exploration project. During the first eighty elapsed days of Rare Metal's permit, the company had drilled 110,000 feet of rotary test holes — the quickest way to determine ore bodies in such large acreages. It was done by simply beginning at one edge of the exploration permit and scraping out a series of roads — 600 feet apart — clear across the entire acreage. Then with six rigs working, holes about thirty feet deep were drilled in the center of the roads, each hole about 600 feet from the next. This eventually gave geologists a pretty fair picture of the entire area. Unfortunately, test holes 600 feet apart can't give an entirely accurate estimate of ore bodies, but prospecting 28,000 acres of rugged canyons and mesas in 120 days doesn't leave much time to speculate.

Always frustrating to the geologists was the fact that one hole would send the counter's needle rollicking all over the scale, while a nearby test hole would register complete sterility. Unlike the other more common minerals, there was no rhyme or reason for the location of uranium deposits.

In addition to its methodical drilling operations, prospecting by air also proved invaluable to Rare Metals. Although it was expensive, dangerous work, an aerial prospector could cover the same area in a few hours that might have taken weeks or months to prospect on foot with a Geiger counter. One such aerial prospector, who literally lived out of a suitcase, was geologist Lee Hanson. He spent most of his waking hours in the back seat of a Cessna 180, hedge-hopping endlessly over the Colorado Plateau. With one eye on the formations below and the other on the sensitive needle of his scintillator, Lee presented a far cry from the pulp magazine version of a prospector. For best results, the plane flew about fifty feet off the ground. This altitude gave him a good look at the country, yet didn't decrease the efficiency of his instruments. When the needle on his scintillator showed he was flying over a "hot" area, this exact location was promptly marked on a map for further, close hand observation.

For the men who lived and worked on the Arrowhead Project in Cameron, life was reminiscent of some of El Paso's earlier pioneering days when the first pipelines were being built across the Southwest. The uranium search at Cameron was tinged with enough excitement to quicken the pulse of even the hardest-bitten prospector or adventurer. Here on all sides was the Painted Desert with its great petrified logs strewn over the ground for miles. Here was the turquoise Little Colorado River wandering slowly toward its destination in the Grand Canyon less than fifty miles to the West. This was the gateway to awesome Monument Valley and strange Navajo names on the map like Tonalea, Kayenta, Oraibi and Mishongnovi. This was country where a man could walk for days and weeks and not see another human. In the reddish haze of this beautiful desiccated land a man without a canteen could die of thirst in a few short hours.

And the ironic part of it was that this vast arid world, long considered to be practically worthless, was now giving up an underground treasure worth countless millions of dollars.

It was actually a pretty good life for the men who worked on the Cameron project. There was an awareness among them that they were doing something big and unusual. There was a feeling of adventure, of pioneering, of being the first to do a job like this. And of course, there was always the exciting possibility of turning up a new uranium strike with their ever present Geiger counters.

Headquarters camp for the employees

of the project was located on a bluff overlooking the Little Colorado, and even with its isolation, it was a rather comfortable oasis. There were trailer bunkhouses, each with living and sleeping quarters for eight men. Another large trailer served as a washroom and bathhouse. Most popular spot in the camp, however, was the mobile mess hall which put out gargantuan meals. Even though it was not quite as fancy as Antoine's or Brennan's, the food seemed just as tasty. Meals were served family style, and a typical noonday spread consisted of roast beef, swiss steak, meat balls, mashed potatoes and gravy, corn on the cob, pinto beans, beets, two kinds of salad, hot biscuits and honey, coffee, tea, milk and pineapple upside-down cake. Meals like this disappeared in a hurry around the hungry exploration crews at Cameron.

Communications at the camp were completely uncomplicated. Transportation was mainly by air, and when company officials, employees or visitors arrived, their planes merely buzzed the camp at roof-top level and headed for the airstrip some two miles away. The "office" manager of the camp usually acted as a one-man greeting and meeting committee. As soon as he heard the incoming plane buzz, he climbed into his pickup and arrived at the airstrip about the same time the plane was taxiing up to the windsock. Harold Horst, who was in charge of exploration and estimating ore reserves says, "We had a fine crew there, the cream of the crop. We had thirteen Navajos on the payroll who knew every inch of that country. They all carried

Geiger counters and they really knew their stuff."

As soon as the exploration crews tabulated their estimates of the amounts and locations of uranium bodies, actual mining operations got under way. Much of the ore located was less than thirty feet from the surface, and with such a relatively shallow overburden to contend with, bulldozers removed the worthless earth. Then, as the valuable uranium ore was reached, it was loaded mechanically into trucks and hauled to the nearest mill.

But hauling ore soon became one of the knottiest problems connected with mining uranium on the Colorado Plateau, caused by the very isolation of the entire area. You can discover a promising show of high grade uranium ore, but if your location is far from the nearest road or mill, you're in trouble. You quickly discover that it costs more to get your ore to market than it's worth.

To overcome this obstacle, Rare Metals got permission from the Atomic Energy Commission to build a uranium mill near Tuba City, some thirty miles north of Cameron. Construction was done by El Paso Natural's own crews and in mid-1956, the Tuba City Uranium Mill was completed and in operation. All concentrate produced from the ores processed at the mill was under contract with the Atomic Energy Commission and it was shipped to Grand Junction, Colorado.

Richard McConn recalls that "it was located on the Navajo Indian Reservation about five miles from the little town of

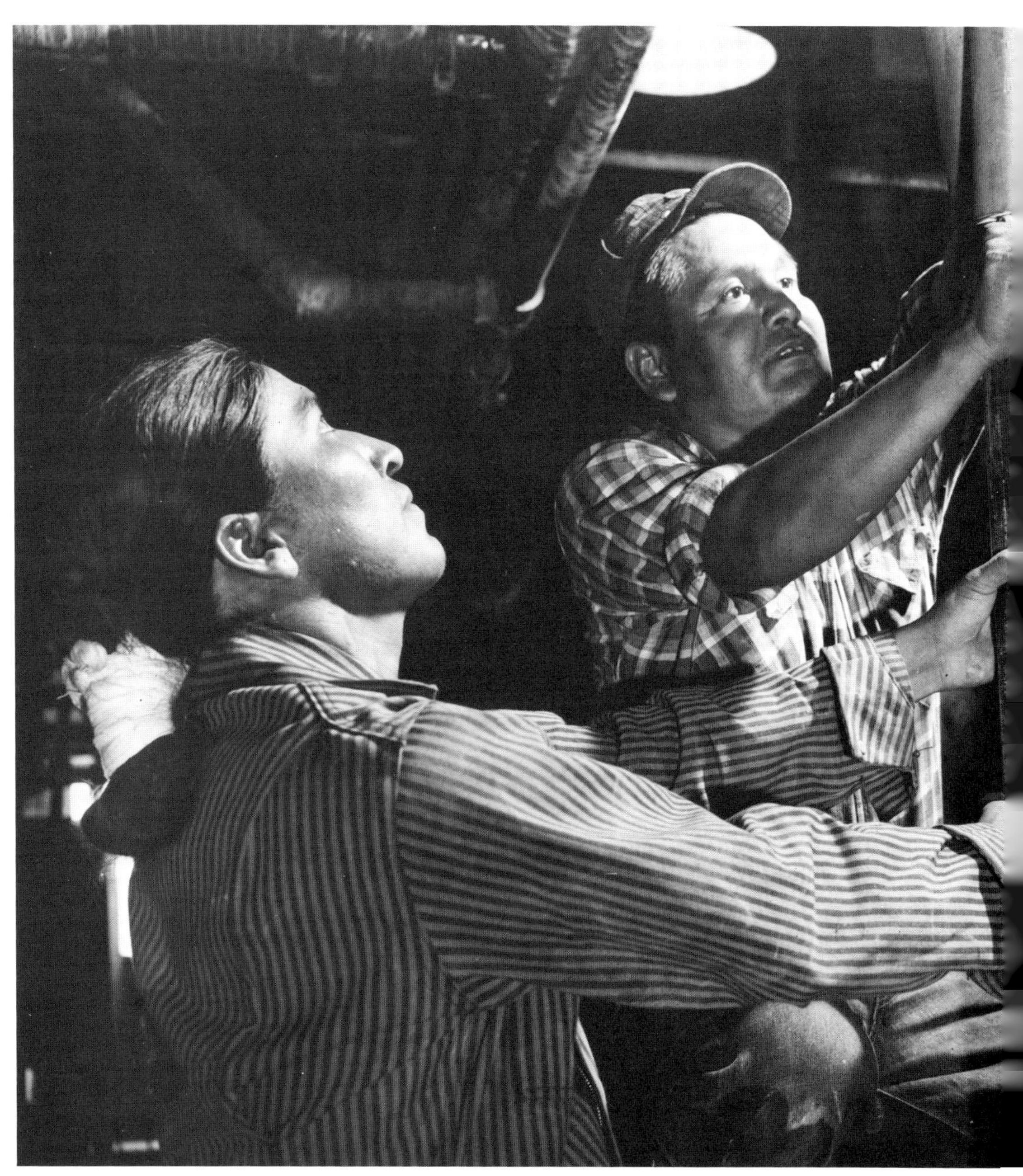

NAVAJO EMPLOYEES at work in Tuba City Mill, which processed all the uranium ore mined at Camero

Tuba City. The town existed mainly because there was a big Indian hospital there that provided medical care to the Navajos for a tremendous radius of several hundred square miles. That's about all that was there, a trading post and a hospital, but a big one.

"All the uranium ore we mined at Cameron went to the Tuba City Mill, so we had a captive source in addition to the fact that it was a custom mill. We bought uranium ores from any other miner that could supply them."

Meanwhile another Rare Metals venture, a mercury reduction plant in Idaho, was moving along at a rapid pace.

The lazy summer afternoon seemed out of character with the hum of activity that was reaching its peak atop Idaho's Nutmeg Mountain, eighteen miles east of Weiser, Idaho. The then governor of the state, Robert E. Smylie, stepped forward and cut a red ribbon at the unloading ramp of the new installation. As the ribbon parted, Rare Metals officially began operating its brand new plant, one of the largest of its kind in the world.

The new mill would process the reddish cinnabar ore the Company was open pit mining on the top of Nutmeg Mountain. The mine itself was not a new one. Long known as the Idaho-Almaden Mine, it was operated by Herbert Hoover, Jr., from 1938 until 1942. However, in '42 the price of mercury plunged to such a low ebb (approximately $1 a pound) that operation of the mine became a losing battle. Mercury prices fluctuated madly for a decade, and in 1954 the price turned upward suddenly and spectacularly to about $4 a pound.

At this point Rare Metals felt that the mine could be operated profitably with the investment of new capital, new exploration and new plant construction. The United States government placed a temporary floor under the mercury price structure offering to buy mercury at $225 a flask (seventy-six pounds), or about $3 a pound. Rare Metals took over the Almaden early in 1955, and by October, the world's largest rotary mercury kiln began processing the cinnabar ore.

"Shortly afterward," McConn says, "mercury began selling for about $400 to $500 a flask. We had a good property up there, a difficult property to evaluate, primarily because the ground was so hard. The rock was so hard it was almost impossible to drill. We wore out bits by the dozens. But at that time we felt we had more than sufficient orders for mercury to build the plant. As history will prove, we did. This was a highly profitable operation for a long time. Then we got into trouble. The Spaniards and Italians control the mercury market. They produce about ninety percent of the world's mercury. The U.S. produces something like two percent. It's not a big mineral in this country. Any time the Spaniards and Italians want to run somebody out of business, they simply lower their price and cut their profits and they do it frequently, so it's a tricky business.

MERCURY ORE is crushed in plant at left, sent by

242

...nveyor belt (center) to mercury reduction mill at right.

STORAGE BINS for crushed ores at the Tuba City Mill are distinctive features of this installation on the Navajo Indian Reservation.

244

We saw mercury go from $500 to $600 a flask down to $125. Even with the U.S. government's price support about all we could do was break even. We didn't make any money, but we didn't lose any, either."

"Eventually," says McConn, "the government support price went off, the market skidded badly and we shut the plant down. Some years later we reopened again because the price came back up to as high as $700 or $800 a flask. We operated again, very profitably for awhile, but the same thing happened. The price fluctuated so badly that we shut the Weiser operation down again, and it has not reopened since. The mercury price is still lousy."

But even with its mercury venture going up and down, Rare Metals had acquired a sound base for exploration, particularly in the search for uranium. In the early Fifties, the race to find radioactive ores raged. Prospectors roamed the hills, poked under rocks with Geiger counters and staked thousands of claims. It was almost too easy. Telltale radiation led right to the ore, and the amateur prospectors with only a Jeep-load of equipment competed successfully with experienced geologists.

Uranium fever was still strong in March, 1955, when Leland Hansen, one of Rare Metals' geologists, arrived at Grants, New Mexico, a small mining town on U.S. 66 midway between Albuquerque and Gallup in the northwestern part of the state. It was a sleepy village until 1950. Then the town acquired an unusual reputation. That spring, Paddy Martinez, a local sheep-herder, was digging postholes in Poison

Canyon, named because of the abundance of loco weed, ten miles west of town. While working, Martinez uncovered a bright yellow rock. Taking it to town, Martinez learned the rock was rich uranium ore. This chance discovery led to the founding of the famous Haystack Mountain mine, operated by the Santa Fe Railroad. Then another mine in the same vicinity was established, and shortly Anaconda Company built a uranium processing mill. Before long, Grants became the booming capital of uranium prospects. The hills echoed to the roar of Jeeps, the clicking of thousands of Geiger counters, the rumble of bulldozers plowing through the fields and the whine of helicopters whirling overhead.

Despite the discovery of vast amounts of uranium, a lot of it was left unclaimed in the late Fifties. Geologists believe the earth is liberally sprinkled with pockets of the ore. Unfortunately, many deposits lie in slender beds under the earth's surface at depths ranging to thousands of feet — a hard place to get at it.

The time had come when surface deposits, once scouted on foot, had to be discarded in the search for deeper deposits. While Leland Hansen was in the Grants area, he went about his job of looking over mining properties. Driving through the town he stopped to talk to prospectors, interviewed mining men and landowners. He sat in cafes and talked to local residents about uranium prospects. One day he visited the quaint old Spanish village of San Mateo, just north of Grants on the slopes of Mt. Taylor. He wanted a closer look at Mt. Taylor, an extinct volcano, active in a recent geologic era. Geologists have learned to follow volcanic activity in seeking uranium. Many believe that volcanic eruptions are responsible for bringing the prized metal to the earth's surface, and in fact, some geologists say the eruption itself is caused by a nuclear reaction.

One thing is certain: Volcanic areas are good uranium scouting grounds. Like other geologists who had visited this area, Hansen was impressed. He believed the Grants and especially the San Mateo area had strong uranium possibilities, but there were other properties in other states he had to examine. After a few days, he left, jotting in his notebook that at the first opportunity, he would return for a closer look.

Before he could get back, another uranium find in the Grants area was made. Twenty-five miles north of Grants is Ambrosia Lake, a shallow, wet-season lake that sprawls across the high plateau. An oil geologist from Houston sank an exploratory hole, and at 300 feet he found uranium. Within weeks, as many as fifty drilling rigs were in operation in a five-mile radius. This quickly became the hottest uranium property in the United States. Mining companies rushed for mineral leases. Soon companies were sinking shafts, making ready to remove the ore and testing out new ore beds.

Returning to Grants, Hansen went out to look at a tract in the San Mateo district which had been submitted to the Salt Lake City office of Rare Metals. One thing bothered the Rare Metals staff. These tracts had been turned down by several other mining companies because they were too far away from the Ambrosia Lake area. But Hansen recommended that the San Mateo tract be obtained. C. L. Perkins, President of Rare Metals, and Mitchell Kline agreed. Drill holes were started, reaching more than a thousand feet underground. Samples of the drill cuttings were taken every five feet and core holes provided samples which were carefully studied and analyzed at Rare Metal's laboratory in Salt Lake. On the twenty-first drill hole sunk at the San Mateo tract, the rotary pushed through seventeen feet of good uranium ore. It was a major discovery.

Working from the discovery hole, drilling crews outlined the main ore body at San Mateo. It was almost a mile long with an average thickness of six or seven feet. It was 250 to 350 feet wide. After learning there was enough ore underground to warrant sinking a mine shaft, workers began excavation on the mine December 21, 1958. Blasting every foot of the way for almost two years, hardrock miners finally pronounced the mine shaft ready for production in the fall of 1959. To the accompaniment of a steel cable whining over its hoist drum, the first ton of uranium emerged from the San Mateo Mine. But there was no time for celebration. Limited production and further development of the mine had to be continued. By early 1960, the mine was producing about 100 tons of uranium ore a day and the company expected a daily production of about 400 tons by the end of the year. Mitchell Kline predicted that

the mine would probably maintain that rate for the next five years.

Previously, during the sinking of the San Mateo mine shaft, there were delays of several months because of great quantities of water encountered underground. Geologists had learned from other miners' experience that the shaft would pass through several water-bearing strata. Uranium ore in the mine was near the bottom of a water-laden formation known as the Poison Canyon sandstone. Below this, was a thick layer of waterless shale, and engineers designed the mine so the main station would be below the ore horizon to escape as much of the water as possible. But, in spite of all that could be done, water flowed from the sandstone to the uranium development areas. In some places in the mine, water dripped from the roof and sides like perpetual rain.

McConn recalls that they were pumping water out constantly so the miners could work. "It was messy," says McConn, "because the ore formation was in a clay type of material, and clay really expands when it gets wet. We put in heavy timbers to hold the walls back, but the clay would expand and you could see the timbers that were maybe six inches square and twelve feet high crack like matchsticks. It wouldn't cave in on you. It would just crack and then the crew would come in and pull that splintered timber out, dig the clay from behind it, put in new timber and it would work for awhile. We were replacing timber constantly.

"The water just literally poured in all the time, and we could never cut it off. I re-

HEAD FRAME for the San Mateo Mine shaft provides necessary strength for the cable and cages as they are raised or lowered in the shaft.

member a funny story about that, though. We were using a procedure called 'grouting' to try to cement off a water-bearing formation. What we did was drill holes into the formation and force cement under tremendous pressure, and lots of it, down the hole hoping we could get enough cement in there to seal off this water.

"Well, about three miles from our mine and downgrade quite a bit, was a small uranium operator who had a fairly productive underground mine. He was probably mining about twenty tons a day. We got a frantic telephone call from him that we were flooding his mine with concrete. It went that far; apparently the water-bearing formation was so badly fractured, it went all over the county and we were pouring tons of cement into this poor guy's mine.

"But the grouting never did work. The water just kept coming into the mine and all the pumping in the world wasn't going to keep it out.

"Then, one cold March day in 1960, all hell broke loose. The hoist man called the superintendent that there was some movement in the shaft. Any time that happens you start looking out. Fortunately, we had some warning. Except for a skeleton crew, we immediately got all the people out of the mine when the movement was detected. The remainder of the men were removed through an escape hatch. And about six hours later, the mine collapsed. With the collapse, tons of material poured into the shaft. But timber falling from every direction formed a bridge about 200 feet from

UNDERGROUND miners use Geiger counter to

248

uranium in San Mateo ore bed.

the surface and the rest of the material fell down on top of that and didn't go any deeper. We had, of course, unbelieveable damage underground, not caused from the things falling from the shaft, but the fact that it cut all of the pump lines and the mine filled with water — completely.

"So we had an underground mine with several miles of working tunnels filled with water."

The debacle at San Mateo was a crushing blow for Rare Metals. The operation was completely shut down and it took months getting all the debris out of the shaft to reopen it and go down with heavy equipment to pump out the water and salvage as much mine equipment as possible. Hundreds of thousands of gallons of water were pumped out, the shaft was cleared and production resumed for more than two years before the company decided to call it quits on San Mateo. But the mine still had one thing going for it. That was its federal government allocation. It meant simply this: uranium mill capacity for ores that were being mined was very tight, and the only way miners could get their ores processed was to have a special allocation from the government. Because of its large size, San Mateo had a good allocation, and Rare Metals approached one of the mill owners and asked them if they would like to buy the San Mateo uranium mine — with its allocation. By this time, the mine itself was not worth much, but the allocation that went along with it was worth a lot. So the mine was sold to United Nuclear, which had a uranium mill in the Ambrosia

Lake area. Rare Metals was able to get most of its investment back through this transaction. According to McConn, "we lost some money, but it could have been a lot worse."

And things did get worse. Rare Metals stock, which had been selling for $10 a share plummeted to about $1.50. The uranium properties at Cameron became exhausted about the same time the shaft collapsed on San Mateo, which everybody knew was going to cost a bundle to reopen. The mercury market declined to a new low, and "suddenly," says McConn, "we were hit by everything at once. You name it and it happened. The government was getting pretty well stockpiled with uranium and the demand dropped. Even today, nuclear generation is nowhere near what it was expected to be back in those days. This is largely because of environmental groups that have held back the government on development of nuclear power.

"I can recall at one time in the mid-Fifties, they were predicting that by 1975 nuclear power would be contributing probably twenty-five percent of all power generated in this country. Now it is something like two or three percent. It just never developed the way that everybody thought it would."

So everything cratered about the same time. Rare Metals went out of existence in June, 1962, when it was merged into El Paso Natural Gas Company. Its activities, from that point forward, were handled by the Mining Division of the Gas Company.

In the large scheme of things, Rare Metals ultimately had a plus factor — the expansion of El Paso Natural's Mining Division. "And had we not been in the mining business," says McConn, "we probably would never have known about the big Lakeshore copper mine in Arizona. Our first interest in Lakeshore was that it appeared to be a reasonably sized open pit operation with reserves of about a million tons of ore. What we ultimately ended up with was millions of tons of high grade copper ore.

"The Lakeshore thing was interesting in the sense that it was owned by a company that was in receivership. In fact, I was over there trying to negotiate with a bank to get the company out of receivership so that we could buy the property. The bank held all the paper on the company at this point because the company owed the bank so much money. The day I was supposed to close the deal, I found out the bank had made a deal with somebody else. That somebody else was William Morley, Chairman of Narragansett Wire Co. So I went up to Rhode Island to see Mr. Morley at Narragansett. Since I knew he was just looking for a copper supply, I told him, 'I know you don't really want to go into the mining business, but we're already in it and we ought to make one heck of a team.' So we wound up acquiring the Lakeshore property jointly. Ultimately, to the advantage of both companies, we wound up acquiring Narragansett Wire Co. And it's been a profitable subsidiary ever since. We

MINERS work on copper ore body at Lakeshore Mine near Casa Grande, Arizona.

sold a half interest in Lakeshore to Hecla Mining Company for one million shares of their stock, which at that time was worth $31 million, which is far more money than Rare Metals ever spent in it's whole existence.

"As far as the Lakeshore copper mine is concerned, at the time we never dreamed what was really out here. We didn't have the slightest hunch that there could be a bonanza. We just thought, here's a property that maybe we could make some money on for the Mining Division. We were only looking at the area they had already drilled, which was all near the surface. I would imagine that the area we are talking about was probably the equivalent of maybe a half of one percent of what is there now. It was nothing, it was peanuts, but it was something we could have made some money on. That's where Perkins came into the picture at Lakeshore. He said, 'Maybe we ought to put down some deep holes, and if we can get the money, let's do it.' Well, we put down some deep holes and we didn't find anything but high grade cooper in every hole we drilled. Perk deserves a tremendous amount of credit for it. It's really his discovery as far as I'm concerned. You might call him the father of what we now know as the Lakeshore Property."

So they closed the books on Rare Metals —in many ways an unlucky adventure that literally drowned in a sea of underground water—tinged with red ink. But as McConn says, it put El Paso in the mining business which could, in the future, lead to bigger and better things.

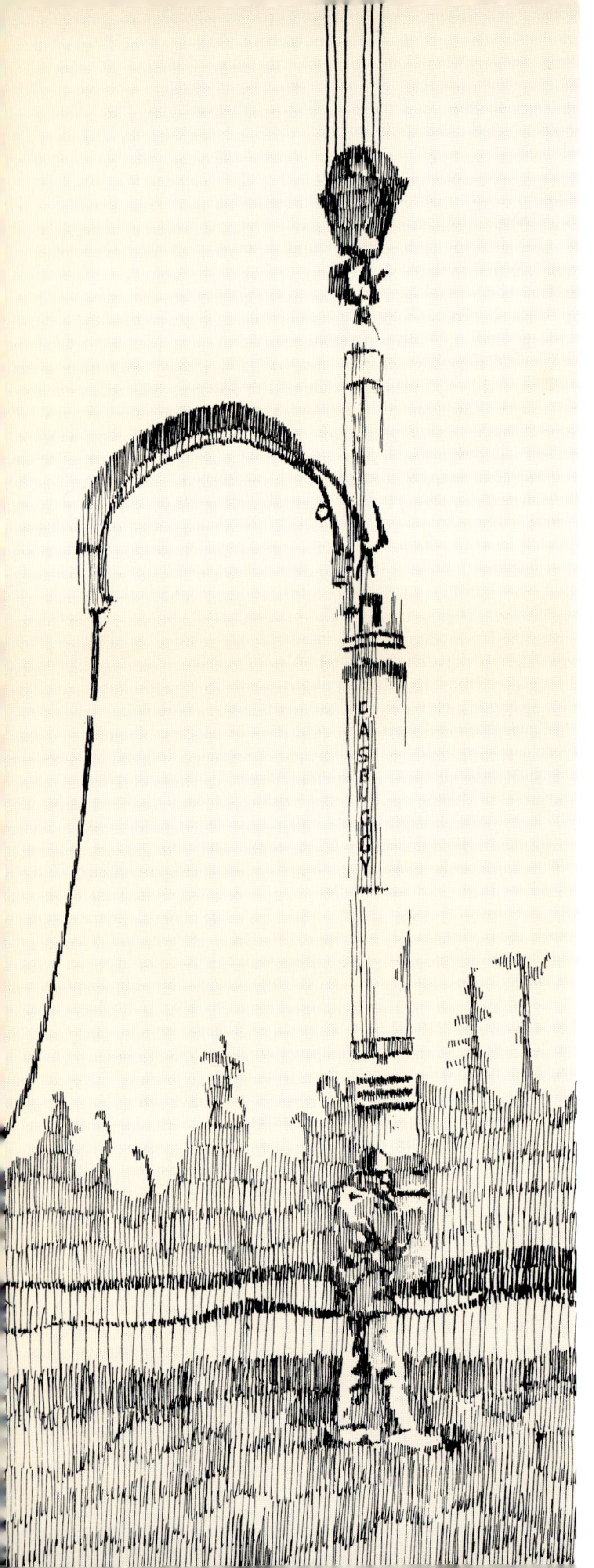

12

TROUBLED
WATERS

A decade of turbulence

NUCLEAR DEVICE
is lowered into the
Gasbuggy well in 1967.

FOR SOME THIRTY YEARS after its founding in 1928, the fortunes of El Paso Natural Gas Company spiraled upward at an almost unbelievable rate. Nothing had been easy, yet new pipeline projects appeared on the drawing boards before compressor stations, plants and pipelines, then under construction, were finished.

But by the late Fifties, philosophies in the natural gas business underwent a drastic change, and a number of incidents took place that were to have a profound effect on El Paso's future.

During the 1950s, El Paso was the sole out-of-state supplier of gas to California and the California utilities began to look for additional gas from others or by entering the transmission business themselves.

This resulted in the entry into California of two new interstate pipeline systems. One, Pacific Gas Transmission Company, controlled by Pacific Gas and Electric Company, brought Canadian gas from Alberta to the San Francisco market area. The other, Transwestern Pipeline Company, transported gas from the Permian Basin fields of West Texas to the southern California area.

Certificated by the Federal Power Commission in 1958, Transwestern entered the California market in 1960 and sold its gas to a subsidiary of Pacific Lighting Corporation. Although Transwestern's thirty-inch pipeline, when operational, was designed to transmit 650 million cubic feet of gas a day, the initial certificate called for moving only 300 million a day, with the company attempting (on a "best efforts" basis) to supply an additional fifty million cubic feet a day.

In El Paso's view, Transwestern would not bring any *additional* gas to the California market since it obtained gas from the Permian Basin, the same source of supply to which El Paso was already connected. What Transwestern did, however, was to increase the price of gas several ways. It contracted to pay 21.8 cents per thousand cubic feet, which was finally reduced eight cents per thousand at the time the certificate was granted. This was a seven cent increase over the price El Paso was then paying producers in the Permian Basin. To meet the competition of Transwestern, El Paso increased the price it was paying producers to seventeen cents. It also eliminated what is known as the "favored nations" clause it had in its various gas purchase contracts. (Favored nations clauses mean simply that in a purchase agreement with a gas producer in a specific area, the purchaser promises to give the producer the same benefits of any higher rates he may make in the future with another producer.) The elimination of these favored nations clauses for gas produced in the Permian Basin and the Hugoton-Anadarko area of the Texas Panhandle, eastern Oklahoma and Kansas, caused a price increase to California utilities of about five cents per thousand cubic feet across the board.

Gulf Oil Corporation was the principal

force in the formation of Transwestern, owning over one-third of the venture (Transwestern is now a wholly-owned subsidiary of Texas Eastern Transmission Corporation).

In 1960, the FPC certificated Pacific Gas Transmission Company. This new organization started deliveries two years later, bringing 450 million cubic feet per day to northern California as well as transporting 150 million cubic feet per day for El Paso which purchased the gas from fields in Alberta (East Calgary and Jefferson Lake) for consumption by the Company's Northwest Division customers. Pacific Gas Transmission's system entered the United States at Kingsgate, British Columbia.

Most of El Paso's Canadian gas was dropped off at Spokane and at various points along Pacific Gas Transmission's line before it entered California.

The net result of the entry of these two pipelines into California deprived El Paso of any future expansion for the number of years it took to blot up the California market demand. El Paso was not destined to deliver any additional gas to California until 1966.

During this period of relative inactivity, El Paso proposed a new pipeline—the Rock Springs Project. This involved the transportation of 400 million cubic feet of gas a day through a thirty-four-inch pipeline from Rock Springs, Wyoming through Utah connecting with systems in southern California at a point southwest of Las Vegas, Nevada. About half the gas was to be purchased from Colorado Interstate Gas Company and the rest would come from El Paso's reserves in various Rocky Mountain Fields and the San Juan Basin.

The FPC certificated the project contingent on approval by the California Public Utilities Commission. In its review of the case, the California commission took the viewpoint that Pacific Lighting (the parent company of the distribution companies in southern California) would be bringing in too much gas too soon and placed a number of conditions on the FPC certificate.

"In light of this," says El Paso Vice President E. G. Najaiko, "the FPC reopened the proceedings to decide again whether or not the Rock Springs Project should be certificated as being in the public interest.

"At this point, one of the country's major pipeline organizations, Tennessee Gas Transmission Company, intervened in the case. Tennessee was a tough competitor. They were selling gas in the New York market in competition with Texas Eastern and Transco, and had engaged in a battle with Texas Eastern to serve the New England market. They also entered the Chicago market, and one result was that antitrust charges were made against the chief executives of Northern Natural, Peoples Gas and American National.

"Tennessee now had its eyes on the California market. It put together a project to bring gas from the King Ranch in South Texas and from Monterrey, Mexico, to use in the boilers of Southern California Edison

254

and the Los Angeles municipal electrical distributor, Los Angeles Water and Power Company.

"The industry promptly dubbed Tennessee's project the 'Enchilada Inch' since it involved bringing 450 million cubic feet of gas per day through a thirty-six-inch pipeline across northern Mexico, entering California at the Mexico-California border.

"Tennessee and the electrical utilities hoped to avoid certain aspects of FPC jurisdiction (primarily the cost of gas at the wellhead) by going through Mexico. They employed a lawyer named R. Clyde Hargrove who did a masterful job of delaying El Paso's Rock Springs Project—sufficiently long for the Tennessee group to put together their Enchilada Inch. The original hearings took only twenty-four days, but the rehearing took over 100 days. In fact, while these were before the FPC, the California Public Utilities held forth on the Tennessee proposal."

At the conclusion of the California PUC hearings, which took some seventy-two days, arguments on the Enchilada Inch took place before the FPC. In the end, the FPC Staff moved to dismiss Tennessee's application. The Staff motion was based on the grounds that direct sales of gas to large electrical utilities was against the public interest, depriving small distribution companies from obtaining gas through their normal supply channels. The FPC Staff also found that projects such as this were merely a device to evade regulation of wellhead prices.

By this time, the Supreme Court set a very important precedent. This was in relation to the so-called "Transco X-20" case in which Consolidated Edison of New York bought gas in Texas, and Transco agreed to transport the fuel to Con Ed's boilers via its Texas to New York pipeline system. The Supreme Court held that this arrangement was also not in the public interest and invalidated the certificate already issued by the Federal Power Commission. One of the principal reasons was that these large electric utilities would use their economic clout to buy gas on an unregulated basis in the field, again denying gas to the small distributor who depended on its supply from the interstate pipeline companies who had to buy gas at fully regulated prices. The FPC Staff saw no difference between the X-20 case and Tennessee's Mexican project.

Within a month or two after the Staff motion, and before the California commission issued its decision, Tennessee collapsed its Mexican plans and immediately came up with what they called their "All-American Project" which soon became known as Gulf-Pacific.

Under this proposal, Gulf-Pacific (a paper entity of Tennessee) proposed to build a thirty-six-inch pipeline from the Katy and Pledger fields in South Texas near Houston to the boiler plants in southern California. The capacity of this 1,500 mile line would eventually build up to 865 million cubic feet of gas per day, about two-thirds of which would go to Southern California Edison and about one-third to the Los Angeles Department of Water and Power. After this filing was made, the FPC issued an order setting a time limit for any other applications to sell gas to California and invited all those who had such an interest to file.

Transwestern Pipeline Company filed an application to deliver 290 million cubic feet of gas per day through its pipeline. This would use all the spare capacity in its existing thirty-inch line bringing its ultimate daily capacity to 650 million cubic feet. El Paso Natural promptly filed for a project to deliver an additional 250 million cubic feet per day to southern California. This gas would be delivered from the San Juan Basin by merely looping its existing systems. Both the Transwestern and El Paso projects were based on contracts filed with the local distribution company in southern California and were mutually exclusive with the Gulf-Pacific Project in that (under the El Paso and Transwestern agreements) Edison and the Department of Water and Power would be served through the local gas utilities as they had in the past, rather than buying gas directly in the field and by-passing the local gas companies.

In order to provide a quantity of gas to the distributor, comparable to that filed for in the Gulf-Pacific Project, El Paso supplemented its application to deliver an additional 515 million cubic feet of gas per day which would come from the Delaware Basin area in West Texas.

Barry Hunsaker, an El Paso Natural Gas engineer who was later to become executive Vice President of El Paso LNG Company,

spent weeks on the witness stand after El Paso made a counter proposal to Gulf-Pacific. "The California market couldn't take more than one project," says Hunsaker, "and we recognized then that if Gulf-Pacific were certificated, it might spell a death knell for El Paso from the standpoint of future expansion of our system. Once that big fat thirty-six-inch was built down there in South Texas, we could see that their incremental cost of service would be rather small and they would have access to some reserves on the Gulf Coast that we didn't have. We might have had to sit idly by for a number of years while the Gulf-Pacific line was not only filled up, but looped, and who knows how far it would have gone.

"To put it mildly, it was a very hotly contested case before the FPC. During 1963 and 1964, we went through 169 days of hearings and produced some 30,000 pages of testimony. We had almost 1,000 exhibits in the case, and we figured once, that if we stacked just one copy of that material onto each other, it would make a stack fifty feet high. That's how much paperwork was involved in the case."

Najaiko remembers the massive undertaking when El Paso's people working on the case were practically living in Washington, D.C. for two or three years. "From our experience at Rock Springs," says Najaiko, "we realized we had been caught unprepared. We came out of Rock Springs looking like hayseeds from West Texas while the city folks were beating our brains out. In comparison, when Gulf-Pacific came about we prepared ourselves in a much different fashion. We started with a "task force" concept. This entailed having a top guy from Regulatory Affairs, a top guy from Engineering, a top guy from Gas Reserves and a coordinator — at a minimum. These fellows spent most of their time on the case and were able to draw on top talent from the rest of the Company.

"Travis Petty was selected to represent Regulatory Affairs, A. M. Derrick took care of all gas reserve aspects and Barry Hunsaker was in charge of pipeline design from Engineering. I coordinated all the information and worked closely with our legal team. It took up a good deal of our time, but we soon found out that this was the only way to get prepared for a case when you have such formidable opposition."

THE CASE OF THE
MAYFLOWER BUG

Headquarters for El Paso's task force was located on the sixth floor of Washington's Mayflower Hotel. Between FPC sessions and in the evenings, the El Paso group met to discuss the day's progress and make plans for the next day's strategy. Unbeknownst to them, every word of their conversations was being monitored by a cleverly concealed radio transmitter and picked up by the opposition forces in the hotel. This was El Paso's first brush with industrial espionage, and it was discovered by Barry Hunsaker.

"I just sort of stumbled onto the whole thing," says Hunsaker. "We were having a meeting in Tony Dungan's room, No. 633. Dungan was from the law firm that was representing El Paso at the hearings. We must have had a dozen people in there because we had, as is usually the case when Howard Boyd is involved, quite an agenda of things to go over. We had rate case matters; we were in the Fifth Circuit on a rate case problem; we had Gulf-Pacific problems; and Mr. Boyd had just returned from Europe and had something to report to us on our drilling in the African desert.

"The meeting lasted a couple of hours. It was about six in the evening on Monday, April 2, 1962. Some of us—A. M. Derrick, Travis Petty, Ed Walsh and I—realized that the subject under discussion didn't involve us and there wasn't any use hanging around so we decided to go to dinner. We left when the meeting was still going on in Tony's room, and Mr. Boyd was speaking. I was the last of the people to leave and as I walked by Room 639 on my way to Ed Najaiko's room (637) at the end of the hall, I heard sounds like some sort of radio communication coming through the louvers at the top of the door. Being curious, I just stopped and sort of cocked my ear and paid a little closer attention to what I was hearing. And what I was hearing was Howard Boyd's voice!

"I knew damn well that he wasn't in 639 because I had just left him down the hall in another room. It was obvious that we were being bugged. They were monitoring us by a speaker apparatus. When I realized what was happening, I wrote a little note to Mr. Boyd and it said in effect, 'I must see you immediately on a matter of great urgency—privately.' And I went back to Tony's room where they were still in conference and just thrust the note at Mr. Boyd and immediately went into the bedroom, where he joined me promptly.

"I told him that we were bugged and that I heard his voice coming out of the door of another room down the hall, and I didn't see how he could be in two rooms at the same time. We quickly decided that we wouldn't let on that we knew we were being kept under electronic surveillance. So we decided to continue the show as though we didn't realize what was happening. The next thing we did was to hire a private investigator named Tom Levenia. He was a private eye, a real pro who had formerly been in the Secret Service during Franklin Roosevelt's administration.

"He came over to Tony's room with some of his fellows carrying electronic listening devices and they quickly spotted the bug under the coffee table."

The bugging device was an FM transmitter about the size of a pack of cigarettes taped underneath the table. It had an antenna about a foot long that was able to pick up everything being said in the hotel room. El Paso immediately retained a well-known criminal lawyer named Paul Connolly, from the law firm of Hogan and Hartson. The first decision was that several Gas Company people should volunteer to watch the room where the receiving equipment was located. Hunsaker smiles grimly as he recalls that "the El Paso people put it to a vote and Travis Petty and I lost the election."

Petty and Hunsaker then began to maintain surveillance over Room 639 and kept a log of everyone who entered and left. A. M. Derrick and Ed Walsh kept up a similar operation from across the hall in Room 633. They were given a brief school on private eye work. Tom Levenia showed them how to tell when a door was opened down the hall by opening one of their windows and attaching a string to it. The string was sensitive enough to air pressure and when a door would open down the hall, the string would move and show that a door had opened nearby. Among other hints, Levenia also showed them how to tape a door lock so that they could open and close the door to their room without making any noise. After their brief training, and reports to the investigating team, Levenia had a pretty good idea who was doing the bugging. He would flash pictures at Petty and Hunsaker — pictures of people around Washington who specialized in that sort of thing.

The two El Paso executives then took turns looking through a crack in the door of Room 637 to see what was going on down the hall, and they stayed in the room night and day, for three days. Hunsaker says that "we could just stand there and look at these birds as they would come and go. It was really kind of interesting, because by then we knew them by name. There

258

were three principals in the operation, a couple of them were former FBI agents. We would log them in and out, and they looked quite furtive. Before entering their room they would always look up and down the hall and knock on the door with some kind of a code knock. Usually as soon as they were admitted, somebody else would leave with a little case of tapes and go downstairs.

"Later they told us that the next time one of the bugging crews left, one of us was to leave our room and go down the elevator with him. When we came off the elevator we were to tip our hat, because we had agents stationed down in the lobby twenty-four hours a day. They would then fall in behind and tail him. I remember the first time it happened, the leader of the gang (it had been rumored that he was involved in the case where the editor of a Dominican newspaper was pushed out of an airplane over the ocean somewhere between the U.S. and the Dominican Republic) got in the elevator and I joined him. I didn't know whether he was armed or dangerous, but he had me scared for damn sure. As luck would have it, we two were alone in the elevator. When we got to the lobby, I tipped my hat to our agents and they put a tail on him."

By the middle of the week Hunsaker began to have some second thoughts about playing private eye, so he suggested to El Paso's group that they send up help in case some real trouble broke out. Tony Levenia responded beautifully, and says Hunsaker, "they sent up a guy who used to play pro football. He was about six-foot-six and weighed about 300 pounds, the biggest guy I had ever seen in my life.

"We figured the buggers knew all the waiters and we sure didn't want the waiters to know we were maintaining a surveillance in that room. When the waiters would bring up our food, we always changed our conversation to talk about the weather and other trivia. Also, every time they would bring our meals we had to hide this big guy. I remember on two or three occasions he threw himself down behind a bed, and once we tried to stuff him into the closet, but he was too big to fit. We had a couple of narrow escapes.

"On Wednesday of that week, we finally got the U.S. Marshal to move in on the case. That night he served warrants to several of the buggers and then later in the evening we went out with the marshal to try to find the rest of the people to serve them with warrants. After spending half the night going around to bars in Washington, where these people were known to frequent, we didn't have any luck. But we sure felt conspicuous walking into these dark bars, Travis and I, peering at the faces of all the people at the cocktail tables. I guess they thought we were a couple of creeps.

"I had a horrible toothache, and the next day I sandwiched in a half-hour to get a root canal job on my tooth. It was a frantic time and we were both pooped, so that night we went to our rooms trying to get some sleep, and I'll be damned if the hotel didn't catch on fire. Also, somebody had shot somebody else in the hotel. Travis called me and said, 'Barry, get up, the hotel's on fire!' I remember saying, 'god, Travis, just call me again when it gets to this floor.' I wasn't about to get up — and didn't. Luckily, they put out the fire in short order."

A. M. Derrick recalls that he was so tired, he slept right through the fire. "The first I knew about it," says Derrick, "was the next morning when I read a banner headline in the *Washington Post*. In big, bold letters it read: MAYFLOWER HOTEL FIRE KILLS TWO. Also on page one was a large headline and story about the bugging incident."

Within a few days, the buggers were arrested and their trials began. Eventually three of them were convicted, but only one served time. It turned out that a Washington police lieutenant was involved, and he resigned from the force. He was found guilty, but the judge threw the case out of court on a technicality. "It was a wild week," says Hunsaker," a real fantasyland."

THE GASBUGGY DETONATION

Under the watchful eyes of closed circuit television and men in hovering helicopters, a nuclear device with a yield the equivalent of 26,000 tons of TNT was detonated in the San Juan Basin, 4,240 feet below the surface of Carson National Forest. It was America's first industrial field experiment in the use of a nuclear explosive,

Project Gasbuggy. The date: December 10, 1967.

The explosion in northern New Mexico was the culmination of an eighteen-month study undertaken by El Paso Natural, the U.S. Atomic Energy Commission and the U.S. Department of the Interior's Bureau of Mines. Samuel Smith, El Paso's director of exploration, (who later became a Vice President) was in charge of El Paso Natural's participation in the Gasbuggy experiment.

At 12:15 p.m. on that freezing Sunday in December, an announcer picked up the countdown of the detonation. At detonation time the earth at ground zero jarred sharply. A shock wave raced through the earth much like ripples made when a stone is dropped into calm water. Seismographs located in the area registered and measured the impact.

Almost two seconds after the detonation, the ground moved under the feet of the observers (some 500 of them) at the visitors' site. A rumbling came from deep underground. National television newsmen had their cameras set on an improvised seismograph made of rocks stacked at the visitors' site. As the shock wave rushed through the earth the pile of rocks tumbled down. A few hundred feet away an official seismograph recorded the shock with jagged lines on a tape. And the crowd cheered.

After a post-detonation briefing, scientists announced that no radioactivity had escaped from the ground and none was expected. The blast went off as planned. It had been felt fifty-five miles away in Farmington, but no damage had been done. The detonation lasted only one millionth of a second, and within a tenth of a second after the detonation an underground cavity about 160 feet in diameter was formed. Scientists who monitored the delicate instruments recording the effect of the explosion were aware that pieces of the overlying rock in the Pictured Cliffs Formation, above the Lewis Formation (in which the explosive was fired) began falling into the cavity almost immediately. This falling rock eventually formed an underground stone "chimney" about 300 feet tall.

El Paso Natural first inquired about nuclear fracturing in 1958 for its potential as a means of inducing greater recovery and higher rates of production from gas bearing formations of low permeability.

Although gas industry people talk among themselves of "pools" and "reservoirs" of gas, there is no such thing as a huge cave or cavern or cavity under the earth that is filled with natural gas. Gas is usually found under pressure in minute pore spaces in rock with some of it appearing to the naked eye as solid as the concrete in your driveway. These pore spaces are connected to each other by tiny passageways. When a well is drilled into the rock, natural gas pressure causes the gas to flow from one pore space to another, then to the well and up to the surface. Reservoir rock, through which the gas moves readily from pore space to pore space, is said to be highly permeable; rock in which the gas moves very slowly is said to be of low permeability. In some areas the permeability is so low that, although there may be plenty of gas in the ground, conventional methods of production can't increase permeability enough for a well to produce gas at a price that users are willing to pay. In industry terms, such a well is not "commercial". The Pictured Cliffs Formation at the Project Gasbuggy site in northwestern New Mexico has extremely low permeability, so low that conventional means of fracturing were inadequate.

Two principal conventional methods have been used historically for increasing production from a well drilled into a sandstone formation of low permeability; one is the use of chemical explosives for "shooting" the well; the other is hydraulic fracturing—the pumping of fluids down the well under high pressure. In both cases the effect is to fracture the rock surrounding the well bore, thus opening up new pathways for the gas to flow from pore to pore through the rock, and enlarging existing pathways. This is called "stimulation" of the reservoir.

The purpose of Project Gasbuggy was to determine whether a nuclear explosive could accomplish the same results as conventional methods of stimulation but on a vastly greater scale. It was believed (and later proved) that a nuclear explosive would stimulate production from sandstone layers over a several hundred foot area.

Drillback into the chimney created by the explosion began two days after the

1. Natural gas is found underground at depths ranging from a few hundred feet to several miles. To make natural gas available for consumption, a well has to be drilled to the formation containing the gas.

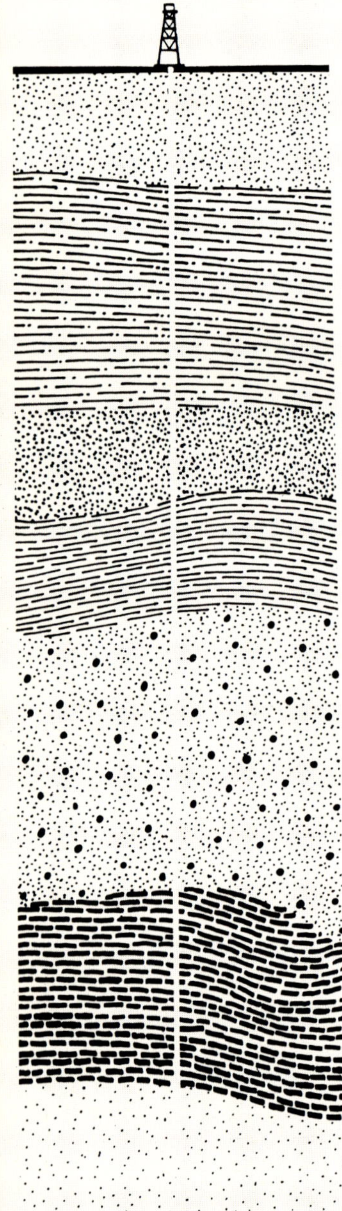

2. Once the well has been drilled into a gas-bearing formation, the gas under high pressure tends to flow through the formation to the wellbore where it can be brought to the surface (produced) and directed into a pipeline. Despite the use of such terms as "pool" or "reservoir" to designate gas-producing underground formations, natural gas is never found in caverns. It occurs instead in rock that frequently is as solid as concrete. In this rock, the gas exists in tiny spaces that in many formations are invisible to the unaided eye.

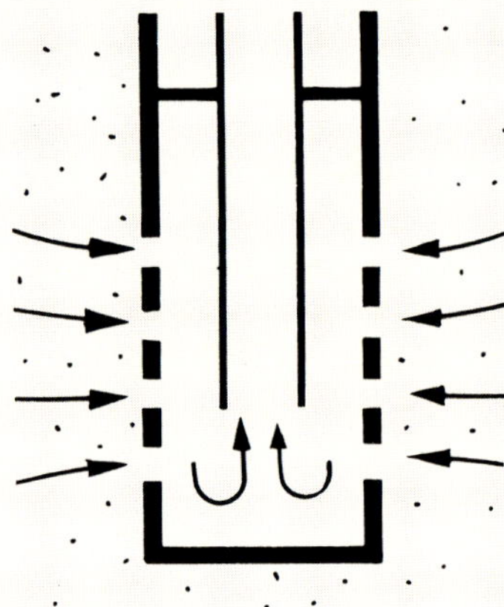

3. In some fields the gas-bearing rock is highly porous and the connections between the pores permit relatively free movement of the gas to the wellbore. A large amount of the total gas is eventually recovered. Illustrated is a sandstone formation as it might be seen enlarged about 30 times through a microscope. Spaces containing gas are shown here in white.

4. Great quantities of natural gas are also contained in very "tight" formations. Frequently, the pathways connecting the spaces that hold the gas are so restricted that the flow of gas through the sandstone is greatly impeded and sometimes entirely prevented.

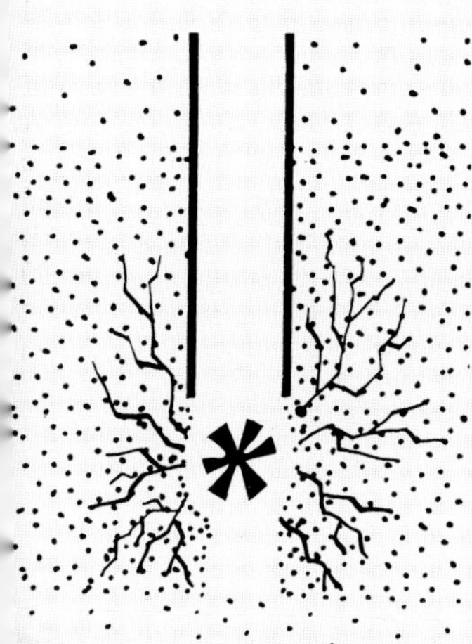

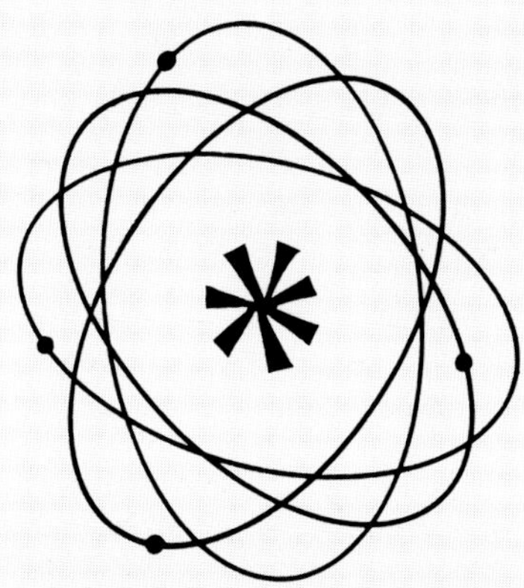

5. For many years, it was the practice to detonate nitroglycerin in these formations, causing cracks which provided passageways which facilitated the flow of gas to the wellbore. This increased not only the rate of flow but the total amount of gas eventually recovered. More recently, fluids under high pressure have been injected into the formations to produce these cracks.

6. Nuclear stimulation is a proposed method of well fracturing which could have great potential benefit. This technique involves exploding a nuclear device in a gas-bearing formation. El Paso Natural Gas Company, the U.S. Atomic Energy Commission and the Bureau of Mines of the U.S. Department of the Interior engaged in an experiment to determine the feasibility of such a technique. On December 10, 1967, the participants detonated the nuclear device for the experiment 4,240 feet underground at a site in the San Juan Basin about 55 air miles east of Farmington, New Mexico. Project Gasbuggy was the world's first experiment using a nuclear explosion for industrial purposes.

7. Indications are that upon detonation, the explosive vaporized the surrounding rock, creating a giant cavity and fracturing the formation in all directions.

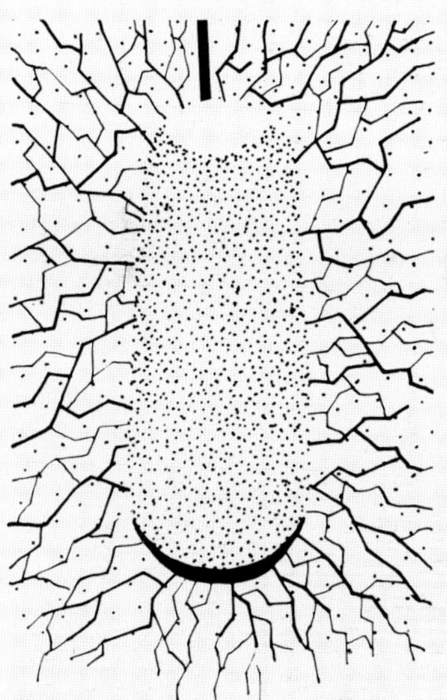

8. Instrument readings show that within approximately a minute, the ceiling of the cavity collapsed, resulting in a rubble filled "chimney." Extensive fracturing extended for great distances through the producing sand, providing many routes for the gas to flow to the wellbore. NOTE: These drawings are not to scale.

detonation and the chimney was reached on January 10, 1968. Scientists took samples of the gas, and the well was shut in for several months to allow short-lived radioactive products to decay to safe levels.

About five months later production testing began to determine how much the rates of gas flow were improved by the detonation and to gather other information needed to estimate the increase in the amount of gas that could be recovered.

A total of 285 million cubic feet of gas was produced from the hole in a seventeen-month period. By comparison, a nearby well completed by conventional means had produced only eighty-one million cubic feet of gas over a ten-year period. The Gasbuggy well was then shut in for pressure build-up observation. All gas flow from the well, except that taken for samples, was flared (burned to the atmosphere) at the site without hazard to workers or to the public.

With public acceptance of further experiments such as Project Gasbuggy, the nation's energy reserves in the Rocky Mountains could be greatly enhanced. Although Project Gasbuggy was a technological success, a subsequent proposal for another El Paso nuclear detonation near Pinedale, Wyoming, never came to fruition. Dubbed Project Wagon Wheel, the experiment envisioned the explosion of five nuclear devices fired sequentially (one right after the other) in the same well bore. Unfortunately, the explosives for sequential detonation were never developed by the Federal Government.

THE PRODUCTS COMPANY CHANGES PACE

In 1958 and 1959, according to Don Thurman, who headed up Petroleum Marketing for the Products Company, the Marketing Department had some 700 people, selling about 7,000 barrels of product a day. "The service station construction," says Thurman, "was handled by our own people. We were building a lot of service stations and there were lots of field construction hands on the Marketing payroll. In addition to that, we were doing all of our maintenance work. We had about thirty people in the property management group acquiring locations for service stations. We also had a lot of salary operated stations, instead of dealer operated units and they were also included on the Marketing payroll. It was pretty top heavy.

"So we had to pare down expenses. And by 1964, we were down to about 300 people on the payroll, selling 40,000 barrels a day of petroleum products. We had 1,450 service stations, eleven bulk plants and two pipeline terminals. Included with the refineries we had about 2,300 miles of product pipeline.

"But the big problem with our operation was simply that we didn't have enough of our own crude production. We only had about 2,500 barrels a day of our own production and all the rest of it was bought at fairly high prices. We had to pay the going rate—$3.30 a barrel for crude.

"In other words, we were caught, as many independent petroleum marketers were, in a cost-price squeeze. We were paying the full price for crude oil and we were having to retail it and keep our retail organization through service stations, through jobbers and through agents. In order to keep them in business, we had to give them a guaranteed margin, and some places the price got as low as twenty cents a gallon — including taxes.

"So here was the Products Company with a fixed price for the crude and suffering a flexible selling price at the gasoline pump — a bad situation."

In the meantime, Lambert Moore, who was general manager of the Products Company, felt that he needed someone to help pull things together within the entire company, a man with lots of experience dealing with people, one who could size up a situation and call the shots as he saw them. That man was Jack Stricklin, who had been holding down the job of sales manager at the Gas Company.

Stricklin puts it this way: "They hired me simply as a Vice President for a reason. I had no specific duty, but it made it possible for me to poke around into anything. You know when you're a Vice President, that means you can do something—you're not limited. And that's what Lambert wanted. I told Lambert when I took the job I wanted to get paid, because it's gonna' take years off my life. So I was workin' every day, and I'd get home at 11:00 at night after bein' in that office all day long.

See, it takes me a long time to soak something up. So I did it with time. Where some people do it in an hour, I do it in seven. And if it happens to be 5:00 that's too bad, but it had to be done. It was a rough time, a very bad time, mentally, physically and everything else.

"Well, I knew they had already committed themselves to some mistakes. To some bad ones. And one of 'em was the motor fuel business. You can't sit there and watch six million dollars go down the drain and not get a lot of ideas.

"One of 'em is to cut your throat.

"The company liked those Red Flame signs, but those little Red Flame signs were givin' me indigestion, that's all I was gettin' from that. They were red, just like the ink that was goin' in the statement. Don Thurman was good, real good, but he was faced with an impossible job."

In 1963, Bill Noel became President of the Products Company, and after reviewing its separate activities, Noel and El Paso Natural's management decided to get out of the retail business as well as the refining end. It was a tough decision, destined to shake up the whole Products Company and its people.

"But," says Stricklin, "Lambert knew Bill Noel was a strong-willed person, and he is. You aren't gonna make twenty-five million dollars and be a pantywaist. You gotta' be tough. Well, the fellas' in the Marketing Department knew he was tryin' to sell it, and as a matter of fact they were out hustlin' it even though they knew it meant their jobs. It was a terrible situation and they'd come in and talk to me about it. And what can you say? A guy says I've got incurable cancer and what're you gonna' tell him, don't worry? It was rough."

So in 1964, Don Thurman was given the assignment to sell the Products Company refineries, service stations and crude gathering system. Ultimately, these facilities were sold to Shell Oil Company, and the decision was made to move all the remaining personnel and offices to Odessa. (That same year, the company's name was shortened to El Paso Products Company.) "Offhand," says Stricklin, "I can't think of anybody who was just enraptured with the idea of moving, especially those who had spent all their lives in El Paso. But me, I didn't care. I'm trained that way. Where's the check gonna' be? I'll be there to pick it up. Everybody was notified just before Christmas that the job was over, and they did everything but have the violins and sad singin'. We'd put El Paso's name all over through six states under our banner and now it was all goin' to Shell."

While the petroleum marketing segment of the Products Company was going through its massive troubles, the Company moved ahead on other fronts that were to prove valuable assets later on. In 1957, the Products Company acquired seventy-five percent of Odessa Natural Gasoline Company (Odessa Natural Corporation). At that time Odessa Natural was operating a casinghead gasoline plant in Ector County. Casinghead gasoline is the untreated, gasoline-like liquid that condenses from natural gas which flows from the casinghead of an oil well. Odessa Natural had just gotten started in the intrastate pipeline business at that time and was supplying the gas for the petrochemical complex in Odessa.

Later, the company acquired the remaining twenty-five percent interest in Odessa Natural, making it a 100-percent owned subsidiary of El Paso Products Company today.

The Odessa Petrochemical Complex continued to mushroom. Bill Noel, President of the Products Company, recalls that, "in 1960, we entered into a joint venture with what then was Rexall (now Dart Industries), to build an olefin plant and a polyethylene plant at the location, and later a polypropylene plant. The original details were worked out, the agreement consummated and commitments made on just a handshake between Paul Kayser and Justin Dart of Rexall. The olefin and polyethylene plants were completed and put on stream in 1962. Shortly thereafter, the Products Company and Rexall combined forces to start a new company, Consolidated Thermoplastics, to produce polyethylene film and a wide range of finished plastic articles.

"The sale of our petroleum marketing facilities was a major improvement for us," says Noel. "This sale enabled us to get completely out of debt and have another platform from which to operate. After this sale we began our next major expansion."

The complex expansion included entries into the fertilizer field and in nylon 6/6 which was a joint venture with the Beaunit Corporation, a manufacturer of fibers and

ODESSA PETROCHEMICAL complex at night.

RESEARCH lab
at Odessa Complex.

textiles, in which the Products Company bought a one-third interest. Beaunit later became a wholly-owned subsidiary of El Paso Natural. The raw materials needed for the production of nylon 6/6 are ammonia, adipic acid, and hexamethylene diamine (HMD). An ammonia plant, a nitric acid plant and a hydrogen plant had to be built at Odessa to supply these raw materials for Beaunit. Beaunit's man-made fibers were primarily used in carpet yarns. In its broad line of knitted textiles, it operated dyeing, finishing and yarn spinning facilities and sold high fashion fabrics to leading garment manufacturers.

A major problem of building and operating an inland petrochemical complex is water supply. Fortunately, it happened that the city of Odessa had a problem disposing of its sewerage plant effluent. The Products Company purchased a large portion of Odessa's disposal problem, treating it and using it (some five million gallons of sewerage effluent a day) in the petrochemical plants. It was a major conservation measure because the effluent had been running through the desert down the Monahans Draw for miles and was nothing but a very contaminated river before the company began taking water out of the draw.

Noel says that "this sewerage effluent is supplemented with another million and a half gallons of fresh water from water leases we have here in Ector County. Another interesting thing about the waste water from our plant is that part of it is now sold to the oil companies in the area to inject back into the ground for secondary recovery of oil. In other words, we make use of our water here two or three times before we are through with it."

So what was once a waste product or by-product of a process is converted into a useful and valuable product.

While things were going smoothly at Odessa, more trouble was ahead, far to the north. In July, 1964, the Products Company purchased phosphate ore reserves and related facilities in southeastern Idaho. The company began construction of a plant near Soda Springs for the processing of the ore into phosphate fertilizers with an initial total investment of about $28 million. Production was scheduled to start in 1965 to manufacture 100,000 tons of fertilizer a year.

"Looking back," says Noel, "the decision to build this plant hurt us pretty badly, and we finally had to sell it. Our timing was bad; we got into it about the same time a number of major oil companies did. Most major oil companies are just like us, they're out of it now."

Jack Stricklin recalls his connection with the fertilizer plant with deep feeling. "I was just tryin' to learn all I could about the new plants at the Odessa Complex and everything was turnin' out fine. No problem. Then Lambert Moore threw his fast-breakin' ball at me. He came in and said, 'Have you been to the fertilizer plant at Soda Springs?' I said, 'No, where is it?' 'Well,' he said, 'it's in southeast Idaho.' So I said, 'Well, Lambert, that's interesting to know;

my geography's improving every day!' And he said, 'I think you better get up there because you gonna' run it.'

"If somethin' didn't run through a pipe, I didn't know anything about it. This fertilizer plant was somethin' else. It's all on conveyor belts and elevators. Gosh, it's a mess if you don't know anything about it. I imagine it's easy if you do. But he said, 'It's a mess and you gotta' go up there and see if you can straighten it out.'

"Well, that plant wasn't gonna' run itself, nothin' ever does. So they're gonna' pull in some yokel like me who hasn't got sense enough to say no. Well, maybe I had sense enough to say no, but not guts enough. So I went up there and I came back and said I don't know what I'm lookin' at, but I don't like it. I don't know who's doin' what, but they were buildin' the plant and makin' a whole stack of work order changes every day, so that meant they didn't know what they were doin' either.

"I went up there just after Thanksgiving in '65, took two days off at Christmas to come to El Paso and that was it. The rest of the time I was up there and worked from four and five in the morning to damn near four or five the next morning, every day. And everything went wrong. You talk about corrosion! If you don't watch a fertilizer plant it will dissolve right before your eyes. Everything was wrong. Almost every place I've been, I've seen people work as hard as the Idaho boys did, but I've never seen anybody work any harder. Their background was different. They were natives of

Idaho, Utah, Wyoming and 'Montana and were phosphate fertilizer plant people — miners, small ranchers or farmers. We, as a rule, were from Texas, Oklahoma and Louisiana and were oil and gas people. Those people up there were great. They're different than we are, but they're wonderful anyhow. And maybe that's what makes 'em wonderful.

"At the Products Company, we had two sets of sales sections. Jesse Owens was sellin' the products we were makin' at Odessa—what we called the organics. Thurman was to sell the rest of it, the inorganics. Basically it was the fertilizer in Idaho. Don set up a good organization, and it would have worked well. We made tons of product at Soda Springs, but the bottom dropped out of the market, and because of the plant cost overruns and fixed charges, we just weren't able to continue operations.

"The next March I went home to Odessa, but I was constantly goin' back up there to Idaho, stay a week or two then come back for a day or two. It was just too much, and finally my heart ripped open. Not open, you know, but I had a coronary and they hauled me to town. And I tried to work after that, but I wasn't doin' a good job for the firm and I felt I was gonna' die if I didn't slow down. I'm not afraid of dyin', but why do it if you don't have to? When you get about my age, you've lost a lot of friends already. And it doesn't make much difference about what you think about an individual. You can like 'em or dislike 'em or just be indifferent to 'em. But every individual plays a part in your life. And when they die a part of your life goes; it leaves a vacuum. Life is just people, that's all. It's not dogs, cats, houses or trees. It's people and that's how it works. At least that's the way I look at it. If I hadn't had this heart damage, I'd have worked 'til they run me off, because I like to. If you can't perform by your own standards you should take the steps to remedy it, and that's what I did.

"In looking back I find that there was little difference between the men in the field and the executives of the Gas Company. They were just country boys with shoes on. Progress is a relentless giant that moves inexorably forward and there comes a time when you are no longer strong enough or young enough or smart enough to move with it and when that happens there are always bright, strong, young men ready to pick up the burden and continue the journey. That is called change and that is good." (Stricklin is now retired and living near Durango, Colorado.)

Almost half a world away from the gleaming petrochemical complex at Odessa, a troubled fertilizer plant in Idaho and the Red Flame signs at Southwestern service stations, the Products Company was out looking for new horizons in its search for oil and gas.

At a landing strip smoothed out of the desert in the Rhourde Hamra area of the Algerian Sahara an El Paso Red Flame symbol lent a bright spot of color to an otherwise yellow landscape. The symbol was on a signboard giving directions to four gas wells drilled by the Products Company's

French subsidiary, El Paso France-Afrique.

The Sahara had long been known as a good place for oil and gas production — even during World War II the United States had sent geological teams into the desert with an eye to future petroleum reserves. Late in 1956, Hassi R'Mel, one of the world's largest gas fields, was discovered in the Algerian Sahara, and Paul Kayser figured many more finds lay under the yellow sands of the Sahara. Kayser asked C. L. Perkins to go over and take a look.

So under Perkins' guidance, Products Company geologists and geophysicists became interested in an area called Rhourde Hamra which was located about 140 miles southeast of Hassi R'Mel. In 1961, the French government granted the Products Company a permit for exploration on one and a half million acres. Two-thirds of this vast area was drifting sand dunes and the remainder was a hard, flat pebbled surface called "desert pavement." Except for the dunes (some were 800 feet high), the area was flat and had none of the characteristic terrain features that indicate subterranean anticlines where gas or oil might be found. The exploration of this area was a job for first-rate geophysicists who could correctly identify underground structures by studying shock waves set off by explosive blasts.

For almost a year, Products people, together with a group of large French banking concerns, roared across the desert in British Land Rovers. Trucks containing seismic equipment and high explosives moved along the paved highway that extended from Algiers through the Rhourde

Hamra to Fort Flatters, an old Foreign Legion post, and the desert echoed with sounds of doodlebugging—geophysical exploration.

A new sound was heard on December 7, 1961, when a French drilling team, working under the direction of El Paso France-Afrique, spudded in the first of the wells in the area. This was wildcatting at its roughest. Although drilling crews ate French cuisine and lived in air-conditioned portable houses, the sand and boredom weighed heavily on them. It was necessary to send drilling crews back to Algiers for two weeks rest after every six weeks on the job. As the summer progressed the temperature skyrocketed until it reached 130 degrees in the shade—and what little shade there was, was at a premium.

But the heat, sand and boredom were worth it. On January 7, 1962, the drilling tempo became tinged with excitement as the first indications of gas were found. On April 1 the well was brought in, and it was a roaring success. Considering that an average of only one well is commercially successful out of every fifty-two drilled, this was a wildcat to remember.

Truett Hollis, who was working directly under Perkins at the time, says, "It was a huge well that could produce about thirty to fifty million cubic feet of gas a day. (This compares with a typical well in the Mesaverde formation of New Mexico's San Juan Basin that produce about a million and a half to three million cubic feet a day; it would take an average of thirty Mesaverde wells to equal the production of one Sahara

well.) And it was on a hugé structure. This
is the one that really kicked off the whole
business over there, because if we had
drilled a dry hole we probably would have
released the concession and got out."

The well was named Rhourde Nouss 1,
and within a month rigs were working on
two more wells. Upon their completion,
a fourth well was begun and brought in.
"Rhourde" is the Arabic word for sand
dune.

"Before long," says Hollis, "we had these
wells all completed and they were just
capped and shut in. But from the work we
had done we knew we had a major gas field
and we really hadn't even tested all the
horizons. We knew there were deeper horizons with possibilities just as great.

"But we were about 650 miles into the
Sahara and our main idea when we first
went in and got these concessions was to
develop some gas and try to move it across
the Mediterranean into southern Europe;
countries like France and Spain and Italy.
The French had done some research work
in laying a pipeline across the Mediterranean."

But fate, in the guise of the long-fought
Algerian Revolt stepped in, and another
Products Company venture faltered be-
cause of timing.

Back in 1954, Algerian discontent with
French colonialism erupted into widespread
rebellion. France had about a half-million
troops in Algeria and they tried, but failed,
to put down the revolt. Rebel guerillas
raided towns, blew up bridges, transporta-

BOWL SHAPED oases in Sahara are called "palmeı

TAUREG from central Algeria.

INSTEAD of leading camel
caravans, many Taureg
workers help to develop the vast
energy resources in the Sahara.

274

tion and communications systems and ambushed French convoys. After years of fighting that claimed the lives of a quarter of a million French and Algerians, Algeria became an independent republic and set out on a program of reform.

Hollis recalls the hectic—and dangerous—days of the revolution while they were drilling in Algeria. "There was a lot of fighting going on," says Hollis, "guerilla fighting, this sort of thing. We used to have to be real careful down there in the desert because the Arabs were going around using plastic bombs in lots of places. They would put a plastic bomb on a wellhead and blow the whole thing off. I remember they did that on a Phillips well over there and the well blew wild for several months before it could be capped. We had a seismic contractor working for us and some of the Arabs stopped the truck one day and killed the driver and one or two other people and stole all the explosives. They were really after the explosives.

"After we would complete a well, we'd put up huge hunks of barbed wire all around just to give us some sort of protection in case they ever did try to put one of those bombs on our wells. As another safety measure, as soon as we completed a well we'd run what we called a blind choke in the bottom of the tubing so that if they blew the wellhead off, this choke would take effect and it wouldn't let the well blow wild.

"So as soon as the Algerians took over from the French we had new bosses on our concessions. Naturally, the Algerians wanted to run their own business and they decided that they wanted to take over and produce our wells. This also occurred on all other production in Algeria. The frustrating thing though, was to get that far from home, do the seismic work and then complete that first huge well — practically all the wells we drilled were producers. I think we only drilled one dry hole. And then to develop what, at that time, could have been about the fifth largest gas field in the world and not be able to do anything about it is pretty discouraging."

But the Algerian Sahara adventure, as discouraging as it was at the time, was later to have its good side. The Products Company's involvement in Algeria had a direct influence on El Paso's entry into the liquefied natural gas business a few years later. Says Hollis, "We knew the reserves were there, we'd proven the reserves. We knew the Algerians. We'd been operating with them for several years. We had offices in Algiers and knew the top people, so when the Algerians took over we could see that LNG was a coming thing."

Hollis spun a globe that occupied a prominent place in his office. "Sometimes when you look at the world printed on a ball like this, it seems pretty small. But when you're walking over the land looking for a good place to drill, the world has a way of stretching out and assuming its real size. The world is big. There is still a lot to be done, a lot of places to look."

The troubled waters of the past decade for El Paso Natural had many employees grasping for the Rolaids. For a company

THE 310 FROM TEXAS: IN 1969, EL PASO
COMPLETED CONSTRUCTION ON A MAJOR
PIPELINE FROM THE GAS FIELDS OF TEXAS
TO CITIES IN WEST TEXAS, NEW MEXICO,
ARIZONA AND CALIFORNIA.

LOWERING IN operation
proved tough going on
steep slopes.

ROLLING CALIFORNIA
HILLS provide background
for this trainload of large-
diameter pipe heading east to
construction sites for the
310 pipeline.

J. N. ROBINSON
(left) chief inspector on
Spread No. 4 checks
progress with Cal
Gaylor as the 310
pipeline winds its way
through the Winchester
Mountains of Arizona.

ROBINSON keeps
tab on entire spread by
using two-way radio.

JOE INGRAM and
Robinson take time
out for a luxurious
lunch break.

RESEMBLING a giant
snake, the 310 pipeline
is gently lowered into
trench cut through rocky
Winchester Mountains.

Hugh F. Steen

that had grown up almost too fast, today's newest project quickly became yesterday's history. Yet the hardships of the late Fifties and Sixties all seemed worth it. Tough new problems tended to wash away any previous mood of self content, and the vexing situations provided a springboard to launch the Company into new ventures.

Top management changes again took place. Howard Boyd was elected Chairman of the Board and Chief Executive Officer on January 15, 1965. The Board announced that Paul Kayser would serve in the newly-created position of Honorary Chairman of the Board of Directors. Hugh Steen, who had been Vice President and manager of Pipeline Operations, was named President of the Company.

While El Paso was busy solving its own corporate problems, the nation went through a period unlike any it had ever encountered. The Sixties wrenched the United States with social and political and moral upheavel that tore the national fabric apart at the seams. And the repairs were to take years to mend.

As forty-three-year-old John F. Kennedy moved into the White House in 1960 a new generation of youth surfaced that was to have a profound effect on manners and mores of the whole country. Four young Negroes entered the F. W. Woolworth store in Greensboro, North Carolina and sat down at the lunch counter for a cup of coffee. Negroes in 1960 simply did not sit down at lunch counters with white folks in some parts of the South, and they were refused service. This episode started a series of sit-ins that quickly spread across the nation, and through protest marches and rallies the resulting black social revolution ushered in a new era in race relations.

Nothing really prepared America for the radical changes of the Sixties. There were signs, if we could have recognized them, in the previous decade: The Beat generation, Elvis, motorcycle gangs, confrontation with Russia, alienated youth and a latent violence and unrest. And most important of all, the U.S. was becoming more and more involved with politics of southeast Asia in a small, divided country called Vietnam.

After the terrible day in November, 1963, when President Kennedy was gunned down in Dallas, one mind-ripping event followed another, and the gentle easy-going days of the Fifties were gone forever. Vice President Lyndon Johnson became President. Hemlines went up as the mini-skirt became a welcome sight for most American males. It was the age of grooviness. Hair grew long and sometimes it was difficult to distinguish boys from girls. There was an explosion of color, and music was everywhere and deafening with an electronically amplified rock beat that drove an older generation up the wall. Drugs moved from the slums to suburbia. Youth confronted its elders with devastating results and neither generation was ever the same again.

America listened to the Beatles, The Monkees, Bob Dylan, The Rolling Stones and hundreds of other rock bands. On television they watched Chet and David with the news, Rowan and Martin, Snoopy and Charlie Brown, Tiny Tim and Johnny Carson. Hollywood's sex symbol of the Sixties was Raquel Welch, whose measurements (37-22-35) endeared her to a generation. Raunchy sex novels dominated American fiction, and books like *Valley of the Dolls* earned author Jacqueline Susann more than a million dollars.

Americans turned science fiction into real life when Neil Armstrong and Edwin Aldrin left their footprints on the moon July 20, 1969. President Richard Nixon commented that the eight-day mission provided "the greatest week in the history of the world since the Creation."

These were some of the outward manifestations of the Sixties—all set against the tragedy of Vietnam.

The war in Vietnam continued to escalate. By 1966, America's weekly bomb tonnage dropped on North Vietnam exceeded the bombs dropped on Germany during the height of World War II. Americans grew increasingly confused as most of them really didn't know when they had started fighting or why. Their dedication became increasingly frayed as more and more of their sons were being drafted and killed. Anti-war protests continued to mount, and finally, on June 8, 1969, President Nixon announced the withdrawal of 25,000 U.S. troops. Thus began the winding down of the most unpopular war Americans had ever known.

Early in 1968, the staff of the Gas Company's executive offices moved into quarters on the top floor of the American General Building in Houston. The offices had been located in temporary quarters in the same

The top photograph shows a city skyline reflected in a pond.

CLEAN, MODERN SKYLINE of Houston (above) changes almost as fast as new photographs can be taken. Above right is the American General Complex where El Paso's corporate headquarters are located. The twenty-four story structure in this picture is the American General Building. In the center is the Riviana Building and at right is the LNG Tower. Picture at right shows the city of El Paso at dusk, where El Paso Natural's operating offices are headquartered in the EPNG Building. Numerous departments are located in other downtown buildings.

building since 1967 when they were moved from Number One Chase Manhattan Plaza in New York. The Company's operating offices were to remain in El Paso, with main headquarters in the El Paso Natural Gas Building, which was bulging at the seams. Numerous departments were scattered in some eight buildings throughout downtown El Paso.

Houston is the center of the nation's oil gas and petrochemical industries, and offers particular advantages as an executive headquarters for companies which are active in these and allied fields. Although the Board Chairman's offices had been in New York since 1957, they were originally located in Houston, where offices have been maintained since 1928. The Company still retained offices in New York for financial reasons.

THE RUSTLERS

EVER SINCE HOLLYWOOD began cranking out one-reelers about the Old West, the great American moviegoer gripped his theater seat and watched the good guys and bad guys blasting away at each other with Colt 45's that shot a hundred rounds without reloading. They fought over water rights, the new school marm and dance hall girls. Greedy land barons and the inevitable cattle rustlers also furnished top-drawer plots. Although there are still a few isolated instances of cattle thieves changing brands and loading up a truck with somebody's prize bull, the heyday of cattle rustling faded into history long before Tom Mix, Hoot Gibson, Gene Autry and Roy Rogers.

But Hollywood may well be missing an academy award by not producing an epic of modern-day rustling—San Juan Basin style. Today's rustlers are a different breed; not interested in cattle, but in El Paso Natural's wellhead mercury and "drip" gasoline. Take mercury, for instance. Recording devices on meters that measure the flow of natural gas from each well contain about ten pounds of mercury. Depending on the prevailing price, El Paso has lost up to $80,000 a year to mercury rustlers. They go through the fields in the San Juan Basin bleeding the mercury from the meters, then selling it to various firms.

According to H. P. Logan, who put in twelve years as division engineer and later as division superintendent in Farmington and is now Vice President in charge of all pipeline operations, it's a tough job to keep tabs on the wild, rugged country in the basin. "If you can visualize it," he says, "there are about 6,500 wells out there and although many of them are only a mile apart, it may be a twenty-mile drive across deep canyons to reach the next well. So you can imagine how much area you have to cover—2,500 square miles. Sometimes there's not a soul around for days except an occasional well switcher and he may have 200 wells to look after. He might get to a particular well every three days, so you can see our problem of catching a rustler in the act."

Posted throughout the basin are posters offering a reward for information leading to the arrest and conviction of anyone guilty of stealing mercury from El Paso's meters. New Mexico State Police have guards on regular patrol in the high mesas and canyons of northwestern New Mexico. Law enforcement groups

NEW MEXICO State Police have had to learn their way around
thousands of miles of dirt road in the San Juan Basin gas patch.

DRIP THIEVES' tools include a wrench, a piece of pipe and a gas can.

have caught dozens of mercury rustlers and continue to maintain a careful watch for others. Halley Ray, who recently retired from the Company, had been chasing mercury thieves for fifteen years as his full-time job.

Logan recalls one incident when Ray caught a rustler stealing mercury between the San Juan River Plant and Farmington. "Ray went up to him, pulled his gun and put handcuffs on him. But he made one mistake; he put the cuffs on the front of this character. Ray turned his back for a moment and the rustler whacked him over the head, knocked him out and stole his Company car. Hank Schulze started calling him 'running gun', and has to this day."

In the late Sixties, when mercury thefts became so widespread, the Company decided on a new tactic. They learned the theory and mechanics of putting a radioactive element in mercury. By "tagging" mercury this way El Paso was able to identify its mercury and quickly locate the middle man or "fence" who bought stolen mercury, and many of the thieves were forced into other occupations.

The theft of "drip" gasoline is another story. As natural gas comes out of the wellhead, it contains a little propane, butane, some oil and other liquids that must be removed before gas enters a gathering line. This removal is accomplished by installing a separator and a dehydrator on each well. The dehydrator removes the water, and the separator takes out the small amounts of hydrocarbon liquids. The remaining fluid is primarily natural gasoline, which according to Logan, "is clear white gasoline and it'll burn in your car just like the product you get at a service station. The gasoline is piped into tanks—some holding up to 500 barrels. Then we go out in the fields with a tank truck and pick up this gasoline, maybe once a week depending on the size of the tank.

"So here's where we have trouble with the drip rustlers. They take a calculated risk that we won't be coming by and they just back up to one of our tanks, connect a hose to the bottom and quickly pump the gasoline right into their tank trucks. They don't sell it around Farmington because it's too risky. The reason is that the state of New Mexico isn't too happy about not getting their tax out of it, so they put a man up there to watch the field, and of course, our same group is watching this too. We had a fellow named Marvin Shaw who used to sit up on a high ridge with binoculars where he could command a view of the basin for miles around, looking for tank trucks that looked suspicious. The drip rustlers used to have a spot down in South Texas that disposed of the stolen gasoline, and they brought a lot of it into Mexico for resale.

"We finally started locking the valves to make it tougher on the rustlers, but these guys would shoot a hole in the tank figuring 'if we can't have it El Paso can't either.' They'll steal anything, even chain link fence. They just roll it up and haul it away in the back of a pickup. Thefts have eased up a lot in the last few years as the risks got greater, but there are lots of interesting things that happen with well operations."

A 157

THE COMPLEX CASE produced thousands of pages of transcripts resulting from scores of hearings, trials and appeals.

13

THE PACIFIC NORTHWEST

A seventeen-year saga of frustration

EL PASO HAS SUPPLIED natural gas to California since 1947, when the state's increasing demands and dwindling supplies caused it to look elsewhere for its rapidly growing energy needs.

Despite the entry into California of two other major gas suppliers in 1960 and 1961, which brought gas in at a higher price than El Paso had been charging, El Paso Natural's low-cost service made it possible for gas distribution companies in San Francisco and Los Angeles to enjoy the lowest wholesale rates of any of the fourteen largest U.S. metropolitan areas.

Meanwhile, in the early 1950s, with California's demand for gas continuing to increase, the Federal Power Commission repeatedly advised El Paso to find new gas reserves to serve the growing West.

About this time, an interesting controversy arose before the FPC between two companies, each striving for authority to bring gas to the Pacific Northwest, the last large area in the country still not served by natural gas.

One of these companies was Westcoast Transmission, a Canadian organization, headed up by a very dynamic character named Frank McMahan. The other company was a Houston based company, ramrodded by an equally interesting man by the name of Ray C. Fish, a prominent engineer and pipeline promoter. He had been responsible for the development of other major pipelines, and he organized a group incorporated under the name of Pacific Northwest Pipeline Company.

El Paso Natural had little interest in the outcome of the case, except for one thing. It was concerned that some of its gas reserves in the San Juan Basin might possibly be diverted to the Pacific Northwest.

At the end of a bitterly fought contest, the FPC favored Pacific Northwest, primarily because it felt that it was not in the public interest to be totally dependent on a Canadian source. So, in 1954, the certificate was issued and construction of the pipeline from the San Juan Basin to the Pacific Northwest was begun.

El Paso's Chairman of the Board, Howard Boyd, recalls, "No sooner had the Pacific Northwest line been completed and gone into operation, it began experiencing financial difficulty. First of all, there had been a large overrun in the estimated cost of the new pipeline. Also, estimates of the demand for natural gas proved to be greatly exaggerated. At the same time, Pacific Northwest had entered into contracts for 300 million cubic feet of Canadian gas which they obligated themselves to take or pay at a ninety percent load factor. Pacific Northwest entered into this Canadian gas contract as a part of the settlement in which Westcoast withdrew its court appeal of the grant of the FPC certificate of Pacific Northwest—which in turn allowed Pacific

DWARFED by Washington's magnificent Mt. R

290

Paso patrol plane inspects the pipeline right of way.

Northwest to finance its project. The distribution companies could not absorb the gas they had contracted for. And had they been held to these contracts it would have probably resulted in bankruptcy for some of these companies and certainly financial distress for others.

"Yet, Pacific Northwest was hardly in a position to give these companies any relief, because it was also in financial trouble. It was not unnatural that Pacific Northwest should propose to El Paso that perhaps there should be some sort of a merger. So, in 1956, we began discussions with them with the idea of merging the two companies.

"Obviously, any discussions contemplating the merger of two companies of substantial size, even though one of them was in financial distress, suggested that we should explore any antitrust implications."

El Paso Natural set out to look at all aspects with experts in the antitrust field. The Company got in touch with Arthur Dean (who was later to become a member of El Paso's Board of Directors). Dean was the top senior counsel of Sullivan and Cromwell, a law firm which had a national reputation in the field.

"It was our conviction," recalls Boyd, "that with the extensive experience we had in the natural gas business, and that with the access to large Canadian reserves, we could turn Pacific Northwest around and make a successful company out of it. We also forecast what has since become a tragic reality, that there would be, in time, a de-

veloping shortage of natural gas in the United States.

"So Mr. Kayser and I, together with Mr. Dean, called on the Attorney General of the United States, Herbert Brownell."

After lengthy discussions the Attorney General felt that since both companies involved in the proposal were fully regulated by the FPC that the merger was not one in which the Department of Justice was particularly interested.

Then the proposed merger with the Pacific Northwest was discussed informally with the Federal Power Commission, which was understandably concerned about the threatened financial collapse of a company it had so recently created by its certification.

El Paso Natural preferred to merge Northwest into itself, a procedure over which the FPC would have jurisdiction. However, Pacific Northwest insisted on El Paso acquiring Northwest's stock from its shareholders.

El Paso ultimately agreed to the stock acquisition route, which was accomplished in 1957 by an offer to exchange shares of El Paso for shares of Pacific Northwest. Later, when the stock acquisition was challenged by the Department of Justice, El Paso felt that all questions related to the acquisition of Pacific Northwest should be resolved by the agency created by Congress to deal with natural gas matters—namely the Federal Power Commission which had control of such mergers. The Company by this time had control of the stock of Pacific Northwest, and El Paso prepared and submitted to the FPC an application for authority to merge the two companies.

After a hearing, the trial judge in a lengthy opinion, set out numerous reasons why such a merger would serve the public welfare. His opinion was unanimously approved by the FPC in December, 1959, which said that the merger was required in the public interest.

After El Paso acquired the stock of Pacific Northwest, over a period of fifteen years it spent in the neighborhood of $170 million (which is a pretty important neighborhood) for new facilities. El Paso tripled sales of gas in the Pacific Northwest and never raised prices to customers during the first thirteen years of operation. In fact, it lowered prices.

In spite of this, El Paso's merger with Pacific Northwest was to create one of the longest court battles in U.S. history. And herein lies the story of a case that has become a textbook example of antitrust — variously described as "antitrust gone mad" and a "comedy of tragic errors."

To fully understand the complex case would require examination of thousands of pages of transcripts resulting from scores of hearings, trials and appeals. Suffice to say, a professor of antitrust laws at Harvard, when asked for his opinion on the suit, put it simply: "The Justice Department guessed wrong when it initiated the El Paso suit."

In 1959, the California Public Utilities Commission appealed the merger, charging the acquisition would decrease pipeline competition in the state by removing Pacific Northwest, which ironically, did not serve California.

Although antitrust laws are designed to create competition, it is an economic fact of life that in the utility field, competition is not desirable. For example, cities have only one telephone company—and certainly subscribers don't want two. There is generally only one electric company, and there's only one gas distribution company. To have multiple companies simply adds to the cost of service. Since utilities are actually handicapped by competition, the substitute for competition is in the regulatory process. This assures customers fair rates, as the rates are limited by a regulatory agency, based on what is considered to be an appropriate profit.

The California Public Utilities Commission was rebuffed by the Circuit Court of Appeals in Washington, D.C., (the only time the broad public interest and needs of the consumer received any examination whatever) but California appealed the decision to the Supreme Court. The Supreme Court then ordered the case returned to a lower court, thus creating what was to become a deep purple cloud covering everything El Paso Natural Gas Company did for many years to come.

The underlying argument for El Paso was the opinion of the FPC which concluded that the merger met the standards of public convenience and necessity and was in the best interests of the American consumer. The Justice Department could have appeared at the FPC hearings, but

EL PASO NATURAL
continued its pipeline
expansion projects in the
Pacific Northwest in the midst
of legal turmoil. Pipeliners
made a 700-foot span across
the Skagit River near Sedro
Wooly, Washington in 1971.

CONCRETE-WEIGHTED
pipeline is laid under the bed
of the Stillaguamish River
near Arlington, Washington in
1970. Completion of river
crossings like this strengthened
the links of the chain of the
long Sumas-Ignacio pipeline
that carried natural gas
across the West.

EL PASO patrol plane
flies the pipeline across the
Columbia River gorge.

IN DENSE FIR FOREST,
welder finishes a section of
30-inch loop line during a
Northwest Division expansion
in 1971.

decided to wait until the California appeal before filing a divestiture order on the grounds that the merger was in violation of the Clayton Antitrust Act "as it could tend to lessen competition."

The Justice Department's first attempt at disrupting the merger ended in failure when the Federal District of Columbia Court of Appeals upheld the decision of the Federal Power Commission. On appeal, however, the Supreme Court sent the case back to the District Court where again, after extended hearings, there were found no violations of the Clayton Act. The District Court approved the acquisition and merger.

At that point it appeared that the case had ended in a victory for El Paso and the consumers, and it seemed to El Paso that the future looked bright. But once more an appeal was made to the Supreme Court. The whole scenario became reminiscent of H. L. Mencken's statement, "There is an easy solution to every human problem— neat, plausible and wrong."

The Supreme Court's ruling was unique. It held that even though Pacific Northwest and El Paso had never served the same markets, the mere "presence" of Pacific Northwest made it a potential competitor of El Paso in California. Without benefit of argument, without evidence presented and without any ruling by the District Court on the point, the Supreme Court in 1964 ordered divestiture — the harshest of all remedies — "without delay."

Dissenting Supreme Court Justice Har-lan said he knew of no other case where the Supreme Court had ordered a remedy not considered first by a lower court. In a review of this action by the Supreme Court, California Public Utilities Commissioner, Thomas Moran, in a scholarly review of the case in 1969, said it was a "legal freak-out" and reminded him of the story of "Alice in Wonderland."

"I am inclined to believe that Lewis Carroll is alive and living in Washington, D.C., where he is ghost writing decisions for the Supreme Court in the El Paso litigation," he wrote.

Following almost endless proceedings, the District Court in 1965 approved a plan of divestiture agreed upon by El Paso and the Justice Department, the only parties to the suit. The chief of the Justice Department's antitrust division described the plan as "the best divestiture agreement in the history of the antitrust act."

However, two companies and the state of California, claiming that they should have been allowed to intervene in the District Court proceedings, filed appeals with the Supreme Court. Again, the Supreme Court decided, for the third time, to hear the case. The issue this time around was one of intervention, not divestiture. But the Supreme Court in 1967, by a divided decision, ruled that intervention should have been permitted and issued a set of guidelines for divestiture. It also removed U.S. District Judge Willis Ritter from the case (Ritter had previously found that El Paso had not violated the Clayton Act

G. Scott Cuming
*Senior Vice President
and General Counsel of The El Paso Company*

and ordered the action against El Paso dismissed). The dissenting opinion of the court described the majority's action as "not only unprecedented but incredible," and "the court roamed at large, unconfined by anything so mundane as a factual record developed by adversary proceedings."

The new trial began in 1967 and ended in 1968 with the selection of Colorado Interstate Corporation as the successful applicant to acquire the properties.

But unbelievably the decision was appealed again, and the case went once more to the Supreme Court which in 1969 sent it *back* to the District Court with new and additional guidelines. The same dissenting judges commented that the action was "dismaying to all who are accustomed to regard this institution as a court of law. They charged that the decision was "to shatter centuries of judicial decision and came from completely erroneous factual premises born of superficial acquaintance with the 14,000-page record."

Twenty-five different parties—including the states of California, Arizona, Washington, Oregon, Idaho, Nevada, Wyoming and Utah; El Paso Natural Gas, the major gas distribution companies operating in California, major gas distribution companies in the Pacific Northwest and in the Southwest, as well as the Justice Department, requested the Supreme Court to reconsider its decision in the case. In the petitions for rehearing, the parties asked that the Supreme Court, for the first time, accord them the opportunity to defend the merits of the plan. The court, however, on June 29, 1970, denied all petitions for rehearing.

A last-ditch effort was begun in 1972 by concerned consumers, distribution companies, governors, public utility commissioners and others to seek legislation to remedy the action of the Supreme Court. Unfortunately, the bill, sponsored by congressional leaders of the Pacific Northwest, was never reported out of committee. Interestingly enough, during all the years El Paso was tied up in these legal proceedings, it had the responsibility to provide service to customers in the Pacific Northwest — and the Company itself never made one appeal to the Supreme Court.

The case was now back in the District Court to attempt a third divestiture decree. In 1973, the judge, in a new divestiture plan chose a group called APCO to receive the properties. The APCO group consisted of the Alaska Interstate Company, APCO Oil Corporation, Gulf Interstate Company and the Tipperary Land and Exploration Corporation. The Supreme Court for the fifth time considered the case, on appeal, and then approved the plan. APCO organized the Northwest Pipeline Corporation, and it was this group which took over the former El Paso properties in 1974.

Howard Boyd, in commenting on the elongated legal battle says the case "illustrates the illogical conflict which prevails today between the antitrust laws and the responsibilities for protecting the public interest vested by Congress in regulatory agencies such as the Federal Power Commission.

"One of the most interesting and ironic aspects of this case," says Boyd, "is that the benefits of the merger of Pacific Northwest and El Paso, forecast by the FPC years ago, all came to pass. As the commissioners predicted, the rates paid El Paso by the consumers of natural gas in the Northwest were lower than they would have been otherwise."

Probably the best summation of the results of the merger were expressed by Howard Boyd in a letter to El Paso shareholders:

"In the usual antitrust case, the public can supposedly expect benefits in the form of increased supplies, improved service, lower prices or all three. In this case, the public received none of these."

But the incredible case was over. The door finally slammed shut on all of El Paso's efforts to provide its service to the Pacific Northwest—after the case went to the Supreme Court five times over a period of seventeen years.

El Paso's adventure in the Pacific Northwest (from the mid-Fifties to the mid-Seventies) was not only a story of judicial proceedings, but of people — like R. W. Harris. "One day, in 1957," recalled Harris, "Hugh Steen called me in and said that we had acquired Pacific Northwest Pipeline Company and he thought I should go up there and run it.

"I went up to Salt Lake City and met with Mr. Kayser and Mr. Steen and took

over the operation without a title or anything. I was using Pacific Northwest personnel and trying to develop a market for
natural gas up there. We transferred and
moved people so we wouldn't have dozens
of job duplications. And it was tough. It
wasn't very long after that that Mr. Kayser
and the Board made me a Vice President,
and that gave me a title and made it somewhat easier.

"But it was still the toughest assignment
I ever had. Everybody up there looked at
me and wondered what was going to happen to them. It would be the same thing as
Hugh Steen walking into Tennessee Gas;
you didn't know the capabilities of the
employees or who was on your side. The
people up there didn't know who I was or
anything. There were officers, qualified
men who had left good jobs to better themselves, and here they were faced with a
whole new deal. But, almost without exception, every top man who left Pacific Northwest came out better after we merged the
companies. Most of the original people that
had been with the Fish Northwest construction group came into our organization and
they stayed right on and maintained their
own seniority."

After some seven years, Harris was
brought back into the Home Office in
El Paso to head all of the Company's
operating divisions. Vice President Estin
Scearce took over the job of managing the
Northwest Division in 1965. In a little more
than a year, Scearce retired, after working
thirty-six years with El Paso Natural. Then
in 1966, William V. Holik (who is now an
El Paso Executive Vice President and a
member of the Board of Directors) was
asked by Steen to take over operations in
Salt Lake City.

"When I arrived in Salt Lake, about
mid-1965," recalls Holik, "a lot of the
problems faced by my predecessors had
been pretty well solved, like the resistance
of the public to use natural gas. By 1966,
we were getting into the situation where we
were sold out, especially in the winter
months, even though we had increased our
capacity to a billion cubic feet a day. The
problems then began to develop in curtail-
ment of service to customers on peak days—
something we had never really had to face
up to in our Southern system operations.

"During my stay in Salt Lake, we completed a number of other major expansions
of facilities and gas supply in the Northwest
Division. In 1967, for example, we placed
new facilities in service which enabled us to
bring in another 100 million cubic feet per
day from Canada. It eased many of the
problems involved—which were numerous
—and we began to make use of existing
facilities to a higher degree. People who
looked at a map of the Northwest System
were often curious about the design. It had
big transmission pipe in the southern portion, suddenly changed to smaller pipe in
the central portion, then just as suddenly
had the largest capacity on the northern
end. Well, you have to understand all the
background of the original certificate hearings to appreciate all the reasons, but
essentially, the Canadian supply was to be
the source of future growth. Both West-

coast Transmission, the Canadian supplier, and Pacific Northwest, the U.S. purchaser, went through ten long years of famine and tough times before the wisdom of the original design was proven.

"I guess this expansion really started to expose many of the problems of gas supply and pricing between the U.S. and Canada which have since become such a source of irritation between both countries in recent years. Then in 1963, we added another 100 million cubic feet per day of capacity. Finally, in 1972, we took the biggest expansion step and looped the pipeline and added facilities to pick up capacity a full 300 million cubic feet a day more. When I left in 1972, we had become what has to be considered as a major gas transmission system. We were moving more than one and a half billion cubic feet of gas per day.

"But if I had to select one single item that most impressed me about what you might call the Pacific Northwest story, I would have to say the people. But how do you really talk about the people? Do you just remember the good, or the bad, or the funny? I don't think so. You have to talk about feelings, the character involved, the loyalties shown, not especially as individuals but as a group or a whole. But think of it; most of the men, and of course the wives, came from other parts of the country and from jobs in other gas transmission companies. Most were involved in the job of building the company from zero, or the ground up so to speak, as the basic construction workers. Contrary to what is often times the case, most stayed on to form the

William V. Holik, Jr.

"I get more hot water faster and for much less with natural gas."

"Natural gas heat brings fresh air into my home."

"Natural gas lets me broil with the broiler door closed."

Natural gas is the finest, most versatile source of energy in the Pacific Northwest. If you haven't yet made the choice, there's no better time than now.

ADVERTISING AND PROMOTION played a major part in selling people in the Pacific Northwest on the merits of natural gas as a fuel.

WORKING in snow and ice in Canadian gas fields was a way of life for field employees, who faced entirely different problems than did their counterparts in the United States.

DEPLETING GAS SUPPLY spurred efforts by El Paso to continue its drilling program. This rig was drilling extremely deep near Fort Stockton, Texas in 1969.

UNLOADING pipe at Flagstaff, Arizona.

304

GAS CONTROL CENTER in El Paso, Texas keeps tab on deliveries throughout the pipeline system.

WAHA FIELD PLANT, in Reeves County, Texas, processes and compresses natural gas for delivery into pipelines moving gas westward.

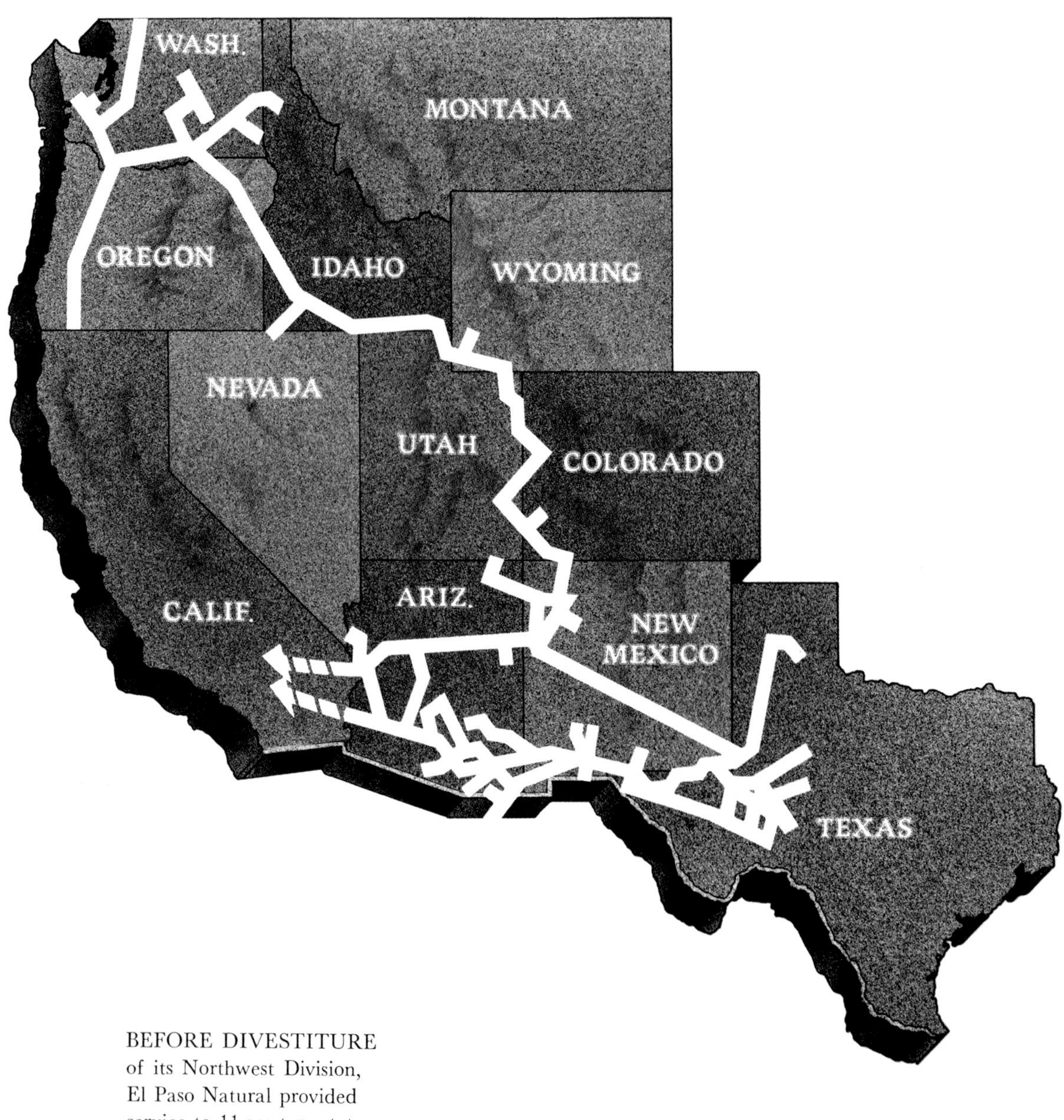

BEFORE DIVESTITURE
of its Northwest Division,
El Paso Natural provided
service to 11 western states.

WRAPPING 34-inch pipe during construction of a loop line near Gallup, New Mexico in 1962.

IN WYOMING, Mark Brown walked from housing area at Station No. 6 to bring his father's lunch to the station each day.

operating group for the new company. And then what happened before they really were able to knit together the organization and establish the long term relationships that make a collection of people more than just that, their leaders were gone and El Paso appeared.

"And when we showed up, along came all the controversy about merger and divestiture, and at the same time, all the problems of building up the business itself. It is really a wonder that people stayed in one place long enough to get their seats warm. But they did, and they worked and did without and got to know each other, and us, and developed the heart and emotion and closeness that makes a real company. Perhaps I'm guilty of being too close to them; I hope I am. But I spoke to R. W. and to Steen about it often and I think they felt the same way. You know, El Paso didn't move in on Pacific and take over everything like many people expected. I don't believe there were ever more than two 'old' El Paso people there at any one time. Pacific people, to use a term, built and earned the success they enjoyed. And I would guess that as they look back on it now, they are in many cases surprised and pleased at what they did. I feel sure they also realize that many of them grew far taller than what they could have guessed in 1956. During divestiture times, when all the different people, both for and against, would visit with me about what could or should happen, I could always say with sincerity, don't worry about making the system go. You have the greatest group of people here that anyone could find.

"However, it finally came to the point of divestiture, and in many respects, it was a relief to everyone. Certainly the cloud had hung over every decision we had to make for the past ten or more years. Divestiture meant not only facilities but in a sense, people also. I guess it's like the story Steen tells about watching your mother-in-law drive your new Cadillac off the cliff — mixed emotions."

Holik returned to the Home Office in El Paso in 1972 and the man he left in charge of the Northwest Division was A. E. McElroy. He had been Northwest Division superintendent for several years while Holik was still in Salt Lake City as Vice President.

McElroy, who is now general construction superintendent for El Paso, says, "I was chief civil engineer for the project when it first got started. It was called Fish Northwest Constructors, and they were acting as agent for Pacific Northwest Pipeline.

"A lot of the employees came from down in the Houston area, but others came from just about all over the country. I'd say that practically every major gas company was represented. To most of them it was sort of a challenge of going to a new area and to a new organization. At the time of the merger, H. P. Logan (later to become a Vice President) came up from Farmington as division operating superintendent, and shortly after that, I was made assistant operating superintendent, working under Logan.

"Within a few years, the system really grew tremendously up there, to where we were serving practically every little town in the entire Northwest. We expanded the system many, many times over those years. It was a far cry from the days when we first went up there with the Fish organization and we'd see big billboards advertising fuel oil, saying 'Use Fuel Oil—The *Safe* Fuel'. I recall quite a few times you'd get questions from anyone you'd meet in the Northwest, a taxi driver, for example. If you mentioned that you were with the pipeline company their first question was, is that stuff safe to burn in my house?' They'd never had natural gas, and the only publicity they'd seen was scary. We turned the whole thing around, and people finally accepted it after our advertising campaigns."

During the final days of his stay in the Northwest, McElroy spent most of his time working with people from the new company that acquired the pipeline. He may have felt like a human Edsel, but he helped them set up their organization and showed them El Paso's systems and methods of operation.

On the evening of February 7, 1974, A. E. McElroy went to his office alone, deep in thought. Had all this really happened? Or was it all just a dream? It all seemed like a fantasyland or a grade B movie on the late show—yet there it was, in living color, this was the end of the line. In the dimness of his office, McElroy quietly cleaned out his desk and turned out the lights.

14

THE SOARING

SEVENTIES

The future lay straight ahead

For more than a decade, El Paso Natural's fortunes seemed to be blocked at every turn. Promising new projects became mired down in seas of red tape, and lengthy (and costly) federal court cases dragged on interminably. The good old days were gone forever; it was a whole new world of environmental, financial and regulatory changes.

And yet in spite of a series of frustrating setbacks, El Paso in 1970, was alive and well, and it was still delivering more natural gas than any other gas company in the United States. Deliveries in 1971 reached an all-time high, averaging almost five billion cubic feet of gas daily.

Even after the divestiture of the North-west Division in 1974, El Paso's total assets were slightly more than two billion dollars and its employees numbered 5,740 assigned to pipeline, gas production, liquefied natural gas, coal gasification and petroleum operations. Another 5,452 were engaged in manufacturing and mining operations.

All across America, the radical social changes of the Sixties began an almost imperceptible mellowing. By the mid-Seventies the hippie movement succumbed to the cold, hard facts of life and as many communes died of malnutrition, their members reluctantly began to find jobs in the mainstream of American industry. But the youth revolt had done its job of social and ethnic change. Even the voting age was lowered to eighteen. The drug scene lost much of its lustre, giving way to alcohol—a throwback to the Twenties. The hard driving acid rock of the Sixties took on a new sound; an intermingling of Black soul music with the mellow brass of the Forties. Best described as "funky," it relates to the flamboyant new style of walk, talk and dress of American Blacks. And finally coming into its own was country-western, making superstars out of musicians like Roy Clark, Charlie Pride, Charlie Rich, Lynn Anderson, and Loretta Lynn.

Disaster and supernatural movies packed the folks into drive-ins and the new, smaller movie houses. Films like "The Exorcist" caused people to spill their popcorn and stumble to the nearest restroom for relief. This was followed by extravagant disaster films such as the "Poseidon Adventure," "Towering Inferno," "Tidal Wave," and "Airport '75." In "Jaws" the sight of a great white shark—in awesome widescreen, living color — eating everything in sight, from little kids to dogs and boats, gave pause to millions of Americans in 1975 as they splashed warily in the shallow waters of the nation's beaches.

TV viewers watched a whole new group of sit-coms that lifted the archaic bars on ethnic problems. Archie Bunker became a household word as a result of "All in the Family." "That's My Mama" and the "Jeffersons" helped break racial barriers with fine actors and good copy. The women's lib movement, which continued to create social changes all over the country, got a boost from "Maude," "Rhoda" and the "Mary Tyler Moore Show." However, a pundit's caustic description of the American scene in the Seventies was that it was "steel-belted, polyunsaturated, solid state, buffered, lanolized and ninety percent pure hokum."

In most parts of the country, men's clothing underwent a radical change as men became style-conscious for the first time in generations. The man in the white shirt, black tie, and expensive gray flannel suit, carrying a brief case gave way to the man in a bright shirt, polka dot tie and an expensive colored suit—carrying a brief case. These same men, liberated from years of drabness, watched with some degree of sadness as the miniskirt slowly disappeared, to be replaced with skirts that dropped to knee-length or all the way to the floor. The feminine look of the Thirties and Forties was in, but thanks to man-made fabrics, combined with cotton, the frumpiness of those earlier decades was missing.

As women's skirts got longer, hair got shorter—and curlier. And (it had to happen) middle-aged businessmen were finally letting their hair grow longer as they accepted the long hair of their sons, who looked like they had their last haircuts about the time Kennedy was inaugurated. At this point, their sons promptly went to hair stylists and got theirs cut much shorter.

As an after dinner speaker at various functions all over the West, Steen's craggy features, still-dark hair which was beginning to take on a light salt and pepper look, combined with his folksy humor, caused some audiences in California to refer to him

H. F. STEEN, then president of El Paso Natural Gas Company, became
Dr. H. F. Steen when he received an honorary doctorate of laws degree from
New Mexico State University during the school's 1974 commencement exercises.

PAUL KAYSER (right) accepts award inducting him into the Permian Basin Petroleum Museum's Hall of Fame in Midland, Texas in 1973. Making the presentation is Bert Haigh, chairman of the museum's awards committee.

INDUSTRY PIONEERS were honored at Aztec, New Mexico in 1971 on the 50th anniversary of the state's first commercial oil and gas production. Pioneer Paul Kayser spoke on El Paso Natural's early activities.

as the Bob Hope of West Texas. His one-liners like "My home town was so poor it was the only place that both sides of the track were wrong," put his audiences at ease as well as giving him lots of credibility in his honest appraisals of the gas industry. With perfect timing, Steen pulled out (and still does) dozens of other homilies at the right time for the right occasion: "The best way to get to Clyde, Texas was to be born there"; "It was so small, we all had to take turns being the town drunk"; "If you buy any whiskey in Juarez, be sure and drive slowly across the bridge — it's not to prevent breaking any bottles, it just gives the whiskey a chance to age"; "Integrity and wisdom are essential in the natural gas business — integrity means that when you promise a customer something, keep that promise even if you lose money — and wisdom is not to make such promises."

During his years as President of the Gas Company, Steen became involved in nearly every fund raising and civic activity in El Paso and headed campaigns for the disadvantaged within his own Company and within the community. He has been described as a man who has risen to take a place of prominence in industry, yet has taken time through the years to become actively involved in the affairs of his community and nation.

In 1973, at a testimonial dinner, Steen received "Man of the Year Award" from the El Paso Board of Realtors as well as the "Human Relations Award" of the National Conference of Christians and Jews. County Judge Udell Moore who, with the El Paso County Commissioners, set aside the day of the testimonial as "H. F. Steen Day."

The county judge recalled that he had come to work for El Paso Natural as an attorney before entering private practice. "I kept hearing nothing but good about Mr. Steen," he said, "and in truth I was almost in fright when I met him in person for the first time. But that fright turned to admiration when, after I told him a particular contract was correct and legal, he asked, 'Yes, I know it's legal, but is it right?'

"He didn't want to know exactly what the contract said, he wanted to know if it was morally and ethically right for all concerned," Moore said.

Probably the highest tribute to Steen was given by Paul Kayser, founder of El Paso Natural, who received standing ovations both before and after his poignant tribute to a man he has known for over forty years, as a fellow employee and friend.

"Hugh Steen is a top executive, in part due to his basic understanding of human relations and his rich sense of humor."

In the summer of 1972, Americans had a ringside seat for the Watergate hearings which were being nationally telecast. They watched in awe as, one by one, President Nixon's top advisors became hopelessly mired in a web of half-truths and cover ups. What started out as a rather minor blunder —the break in by Republican party members of Democratic headquarters in Washington's huge Watergate apartment complex — ended up with the American public seeing a jagged crack emerge in the mirror

THE OLD AND THE NEW. The red and yellow EPNG torch symbol was phased out on all vehicles and printed matter beginning in 1972. The new corporate logo (right) reflects the bold, clean graphic look of the Seventies.

of their beloved democracy. After hearing a series of incriminating tape recordings made in the Oval Office, Congress and the general public was ready to impeach the chief executive. Before proceedings could be instigated against him, Richard Nixon took the only practical way out. He became the first president in the history of the United States to resign his office.

To Nixon's credit, however, he managed to extricate American troops from Vietnam, an unhappy war that had its beginnings for the U.S. during the Kennedy administration. He also established a detente with China and Russia that had been thought impossible.

In 1974, C. L. Perkins retired as Senior Vice President and Director of the Company after forty-five years of contributing to the growth and success of El Paso Natural.

In a warm resolution, the Board pointed out, "His exceptional knowledge and judgment, his warm and generous personality and the wise counsel he had given so unstintingly to those in the El Paso organization fortunate enough to have worked with him, have brought great strength to the management of the Company."

These words of tribute echoed throughout the Gas Company, from one end of the pipeline to the other and throughout the various activities Perkins had been instrumental in organizing. Since the Company's very beginnings, he was an innovator, a pioneer in every sense of the word. He quietly set in action new methods and standards for the entire industry with a vitality and an innate ability to get things done. Perkins has been called the outstanding pipeline man in the world today; others

316

Robert J. Feldman
President of Narragansett Wire Co.

IN PAWTUCKET, Rhode
Island, Narragansett Wire Co.
employee moves in-process
copper to manufacturing area.

"The Magic Suitcase"
PRESENTED BY
Natural Gas Company

IN ITS FORMATIVE
YEARS, EL PASO NATURAL
WAS PRIMARILY A MAN'S
WORLD. TODAY, MORE
AND MORE WOMEN ARE
TAKING THEIR PLACES IN
THE CORPORATE
STRUCTURE. THEIR
JOBS COVER A WIDE
RANGE, INCLUDING
MATHEMATICS,
GEOLOGY, SECRETARIAL
SCIENCES, AND OFFICE
MANAGEMENT.

say "he wrote the book." He possesses a toughness necessary to do business in a tough environment, yet this quality is always tempered with a personal, warm, human touch. "Perk" is one of those rare individuals who has become a legend in his own time.

Meanwhile, the cost of living spiraled upward at a frantic pace. Inflation gripped the nation, followed by recession. Top economists predicted that the recession could develop into a full-blown depression reminiscent of the 1930s. Nobody could agree on a cure. But the nation's economic ills were soon to be amplified by a new phrase people began reading in the press: "energy crisis."

For twenty years, El Paso Natural had been sending up warning signals that the supply of natural gas and oil is not infinite. Federal government regulations had kept the price of natural gas so low that the country was using it up at a fantastic rate, but nobody really listened. In 1968, for the first year in history, more natural gas was consumed than was discovered.

Surveys showed that most Americans didn't *really* believe there was an energy shortage. As the world's greatest oil producer the country had grown fat and prosperous, feasting on an abundance of energy. America was consuming one-third of the world's energy, although it accounted for only six percent of its population. Any thought that we might be forced to go on an energy diet was totally bewildering.

"Just another conspiracy of the big companies putting the squeeze on the little man," many said. One Wisconsin congressman was less polished in his allegation. "There is little doubt that the so-called energy shortage is just a big lousy gimmick foistered on consumers to bilk them for billions in increased gasoline prices," he said.

In the meantime, the Arabs placed an oil embargo on the United States in 1973 primarily because of our military and economic support for Israel. And this hit people where it hurt—at the gasoline pumps. They were lining up at their friendly neighborhood service stations impatiently waiting to buy three gallons of gasoline and glad to get it. Teeth gnashed, nostrils flared and blood pressure skyrocketed as the motoring public fought for its place in line.

Hugh Steen puts it this way, "Few people in the United States understand the magnitude of today's natural gas shortage. The shortage is equated by many to a shortfall of oil. Yet, natural gas is the single largest source of energy for residential, commercial and industrial markets. On a BTU basis we use five times as much natural gas as electricity, and gas provides energy for almost fifty percent of our industrial capacity.

"And natural gas is a premium fuel which does not pollute and is far more efficient than any other energy source. It costs far less than any other form of energy, which is really the reason for today's natural gas shortage."

Natural gas in interstate sale (such as El Paso operates) is controlled by the Federal

HOWARD BOYD (center), Chairman of the Board of The El Paso Company, presides at annual meeting of stockholders in Wilmington, Delaware. From left are: John B. Megahan, Secretary; E. G. Najaiko, Vice President; Boyd; H. F. Steen, President; and W. Burney Warren, Executive Vice President and Treasurer.

RESTRUCTURING

Now it's The El Paso Company

IN MAY, 1974, shareholders approved a major restructuring, establishing a holding company appropriately named The El Paso Company.

Because of the many new projects of unprecedented size and nature (foreign to the natural gas business as it had been known in the past) the Company encountered problems when it undertook to use El Paso Natural Gas Company as the vehicle for financing such things as the LNG project, and it visualized difficulties for financing future major projects such as the Iranian and Russian ventures. Since these huge projects entail different risks — in some instances political and technical risks — it was decided to establish a holding company of which El Paso Natural Gas Company would be one subsidiary.

Major affiliates under the restructuring are El Paso Natural Gas Company, whose principal business remains in natural gas operations and domestic exploration; El Paso Products Company, petrochemical operations; Narragansett Wire Co., production of copper wire and cable products; El Paso LNG Company, Algerian LNG projects; El Paso Energy Resources Company, coal and coal gasification and related projects; El Paso Development Company, land development projects; American Corporate Resources, insurance; El Paso Building Company, property management; and other LNG and new energy ventures.

Power Commission. The Commission, in 1954, stated that the price charged by a producer was subject to FPC jurisdiction, and the Commission set area rates for gas at the well head. Because the price was and always had been too low, the producers sell their gas to *intrastate* (gas that does not go out of state) markets. Those markets are not controlled by the FPC. The results have been almost catastrophic for interstate pipelines such as El Paso, which had been unable to make any significant gas buys since 1968. The answer is simple: Intrastate buyers pay more, perhaps three or four times more than El Paso is allowed to pay. So why should producers sell to a company like El Paso?

Compounding the plight of the interstate pipelines are the restrictions on exploration and development, particularly in the offshore areas where the bulk of unfound gas may be. In the continental United States, the "easy" gas has already been found, and wells are drying up every day of the year.

Much of the fault for the failure to open up sizable segments of offshore areas for bidding was the Santa Barbara blowout which was used to great advantage by environmental extremists. In addition to the environmentalist problem, El Paso was also being hurt by the regulatory lag. Literally years are taken before anybody can get a decision on a major energy project. A case in point is El Paso's liquefied natural gas project with Algeria. The Company filed for approval of the project in 1970 and didn't get final confirmation until 1973.

EL PASO'S LNG PROJECTS

Back in 1962, El Paso Natural's exploration crews discovered several huge gas fields in the Algerian Sahara. In the years that followed, both before and after Algerian independence, El Paso was active in development work on these discoveries. The Company has always enjoyed good rapport with the Algerians and in particular those employed by Sonatrach, the Algerian national oil and gas company.

Various possibilities of transporting gas were considered, including piping the gas under the Mediterranean and shipping gas in liquid form to Mediterranean ports. Other projects based on these resources such as the construction of petrochemical plants were considered, and rejected, since in 1963 America's energy demands were not conducive to bringing the gas to the U.S. But there has been a dramatic change in the nation's energy picture since 1968 (from 1970 to 1975 the contiguous forty-eight states consumed more than twice as much gas as had been found there).

In 1968, El Paso and Sonatrach began discussions on the marketing of Algerian gas in the U.S. El Paso agreed to conduct feasibility studies, and the first Company LNG team was formed—headed by George Carameros, and supported by Barry Hunsaker and Virgil Cowart. The following year a contract was concluded with Sonatrach for the purchase of LNG, equivalent

George D. Carameros, Jr.

SIGNING the contract for the importation of Algerian liquefied natural gas to the eastern part of the United States are Howard Boyd, El Paso Chairman (left), and Sid-Ahmed Ghozali, President of Sonatrach, the Algerian national oil and gas firm.

to one billion cubic feet of gas per day. Sales agreements were later concluded with Columbia Gas in Wilmington for 300 million, Consolidated Natural Gas Services Company in Pittsburgh for 350 million and Southern Natural in Birmingham for 350 million. The contracts with Sonatrach and El Paso's customers are all for a period of twenty-five years.

The project, called Algeria I, is designed to deliver one billion cubic feet of gas a day to the East Coast of the United States. About two-thirds of the delivery will be made to a new LNG terminal and regasification plant at Cove Point, Maryland, and one-third to another new facility on Elba Island in the Savannah River near Savannah, Georgia. The LNG will be loaded aboard vessels at a new liquefaction plant near Arzew, Algeria. All of the facilities related to the project, including ashore and at sea, have been specifically designed for this one project.

Approximately twenty-eight Algerian gas wells will be required, and on an average day, each well will produce about sixty million cubic feet. The gas, which is sulfur-free, will be delivered to Arzew in western Algeria by a forty-two-inch pipeline. The 315-mile pipeline will have four compressor stations to keep the gas moving from well-head to the coast.

Here, the gas will be processed into LNG in one of the world's largest liquefaction plants, and liquefied by using the science of cryogenics (super cold) — freezing it to a temperature of 263 degrees below zero

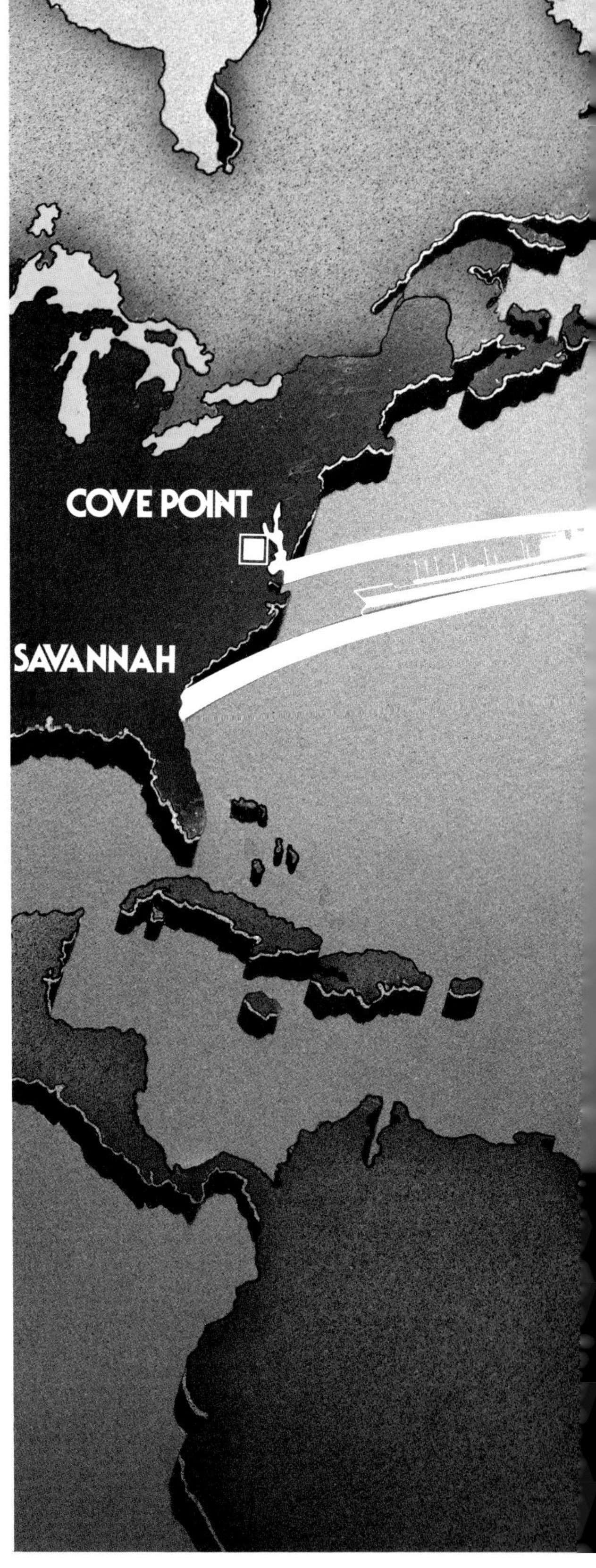

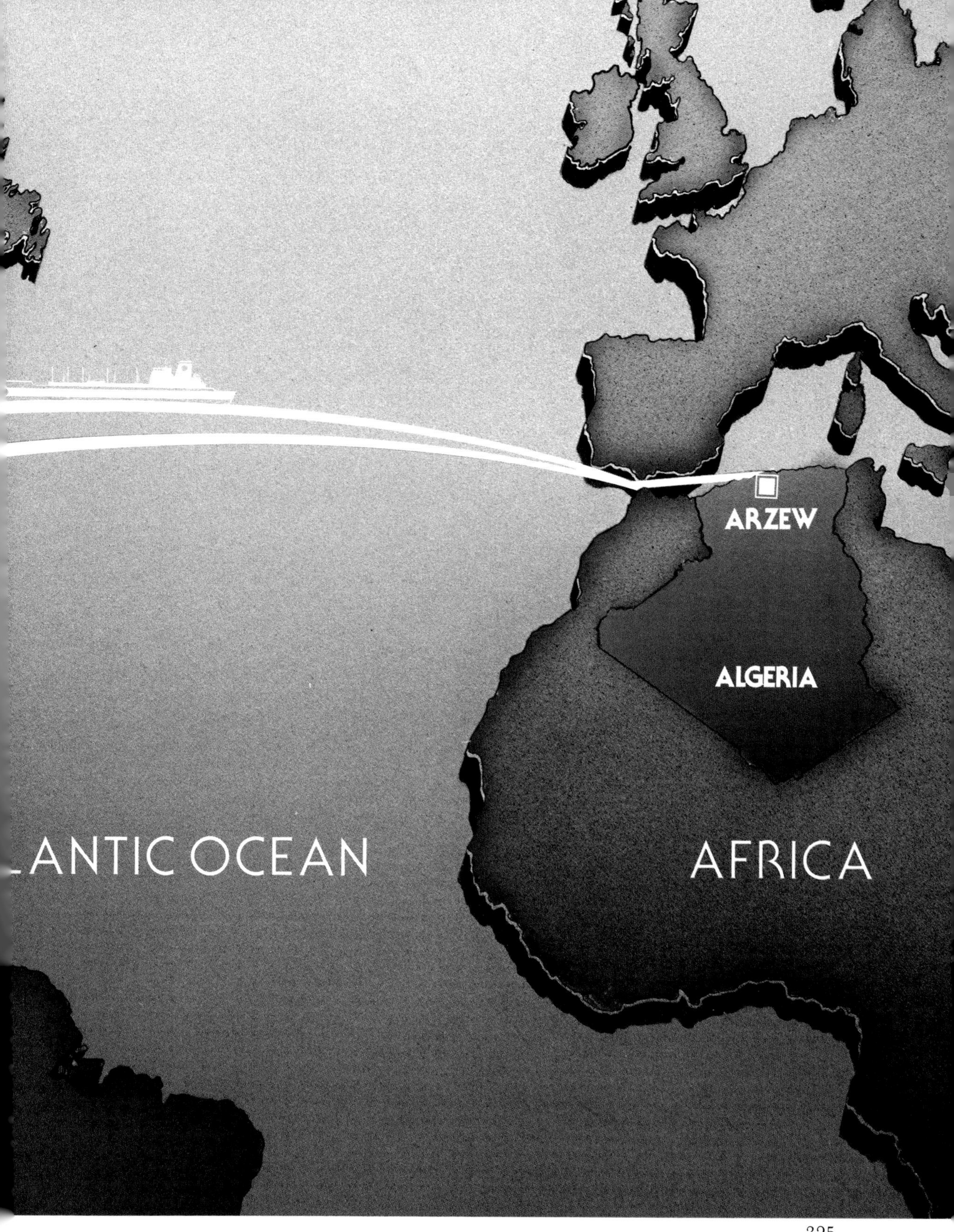

ARZEW
ALGERIA
ATLANTIC OCEAN
AFRICA

Fahrenheit. The resulting liquid takes up only one-600th of the space normally required for natural gas in its original form. The LNG will then be pumped aboard the cryogenic carriers for the voyage to the U.S. eastern seaboard.

The nine carriers required for the project will be, by far, the largest vessels of their type ever constructed, each about the length of three football fields and capable of holding the equivalent of some two and a half billion cubic feet of natural gas. The ships will be capable of sustained speeds of twenty knots, can be loaded and unloaded in about twelve hours and will make a one-way trip in an average of about eight and a half days.

First of the giant ships, aptly christened the "El Paso Paul Kayser," was completed and sea-tested in the summer of 1975. Shipyards in Dunkerque, France are responsible for building the first three ships. Six others are being built in the United States. All of the ships will be completed by late 1978.

George Carameros, President of the subsidiary, El Paso LNG Company, feels strongly that LNG is not going to be just a temporary solution to America's energy crunch. "It is going to be a permanent factor," says Carameros, "in complementing and supplementing the nation's gas supplies in the future. LNG will be making vital contributions to the industry's base load requirements for a number of years, while the results of added incentives to increase domestic gas exploration efforts are awaited and while other technologies — such as gasification of coal and nuclear stimulation—are being developed.

"The immense investments required for pipelines, liquefaction plants, cryogenic carriers, and terminals and regasification facilities make it essential that LNG projects like this be undertaken on a high load factor basis if the delivered price is to be at acceptable levels. The ships, alone, for this one project will involve the investment of approximately a billion dollars."

It is estimated that the cost of all facilities, those in Algeria, the cost of the ships, the facilities on the East Coast of the U.S., will be in excess of $2.5 billion. At the time the project was announced, this represented one of the largest commercial ventures ever conceived by man in world history.

Application to the FPC for approval of a second project to import natural gas from Algeria was made early in 1977. This project is known as Algeria II. Under Algeria II, equivalent to a billion cubic feet of gas a day will be sold to the Company by Sonatrach and landed on the Texas Gulf Coast near Port O'Connor where a terminal and regasification facilities would be constructed by El Paso. Twelve ships will be required, six of the twelve will be supplied by Algeria.

From Port O'Connor, thirty-one miles of thirty-six inch pipeline will be constructed to connect with the pipeline system of United Gas Company and another thirty-inch line will be built by El Paso. This new pipeline will extend westward for

432 miles to the Company's Waha Plant, southeast of Pecos, Texas. Five compressor stations located near Victoria, Seguin, Kerrville, Sonora, and Fort Stockton, will push gas through this line.

It is planned that 650 million cubic feet a day will go to customers in El Paso's traditional service area in the West, and 350 million cubic feet a day will go to United's customers.

LNG CARRIER, El Paso Consolidated, under construction at Chantiers de France Dunkerque.

SONATRACH'S pipeline from Algerian gas field at Hassi R'mel crosses the
Sahara into the Mediterranean port of Arzew.

ALGERIAN walled city of Uzahts.

CHILDREN play at Hassi Messaoud in heart of gas fields.

FIRST OF THE FLEET,
the El Paso Paul Kayser lies in
dry dock in St. Nazaire,
France for painting and
inspection after construction
in Dunkerque.

EL PASO EXECUTIVES pause during a tour in Algeria to examine a scale
model of the natural gas liquefaction facilities in Arzew.

ATTENDING LAUNCHING of the LNG carrier El Paso Southern in January, 1977 are (from left) Hugh F. Steen, Mary Steen, John P. Diesel and Jan Diesel. Diesel is Chairman of the Board of Newport News Shipbuilding and Dry Dock Co.

AFTER CHRISTENING
the El Paso Paul Kayser,
Mrs. Kayser receives
bouquet of flowers from
Gerard Chauchat,
Chairman of Chantiers de
France Dunkerque.

HOWARD BOYD and
H. F. Steen during
ceremonies at Dunkerque.

LAUNCHING CEREMONIES for the El Paso Cove Point at Avondale Shipyard in New Orleans (above). The El Paso Columbia (below) is towed from submersible drydock at Avondale.

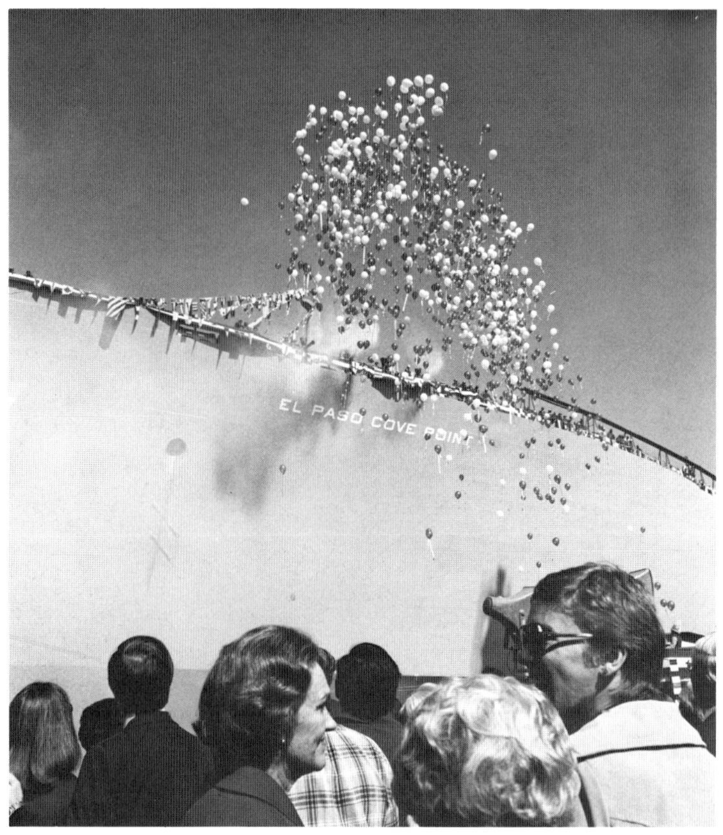

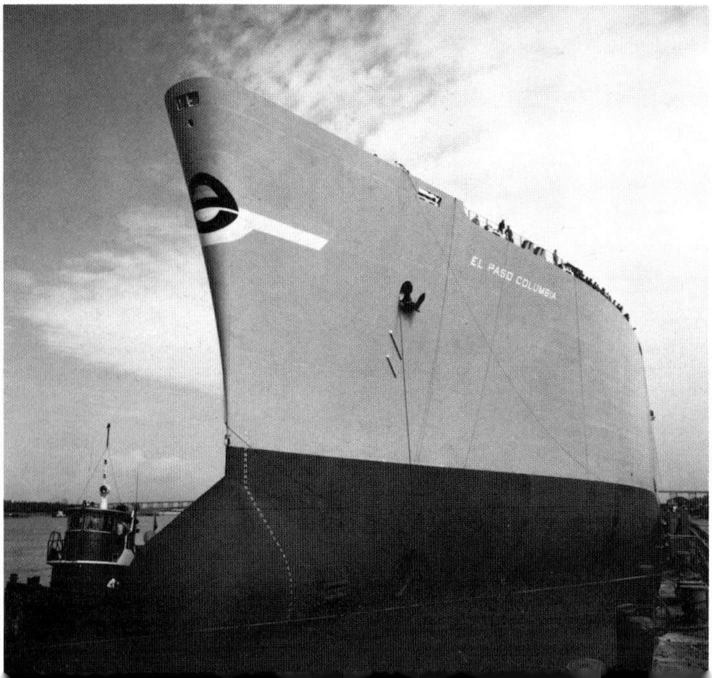

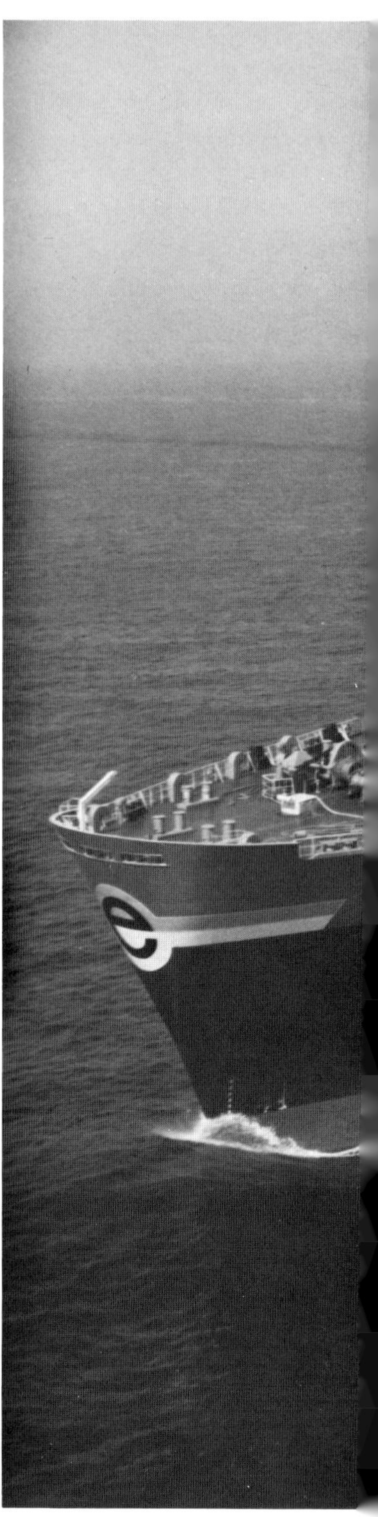

EL PASO PAUL KAYSER undergoes sea trials in the English Channel.

One project that was six years in the making, the Lakeshore copper mine, began taking final shape in the fall of 1975. In late September, project officials threw the switches to mark the start-up of a $192 million underground copper mining and processing operation thirty miles south of Casa Grande, Arizona on the Papago Indian Reservation.

The Lakeshore copper project is a joint venture of El Paso and Hecla Mining Company with Helca as operator. One of the nation's most highly respected and successful mining companies, Helca, with headquarters in Wallace, Idaho, is the nation's largest domestic silver producer. In addition to the extensive underground mining development, facilities at Lakeshore also include surface plants for processing ores from the mine to produce cathode copper and copper precipitates. At full production, the mine and ore processing plants will produce about 135 million pounds of copper a year.

However, the copper industry in 1977 faced severely depressed prices. Before any contribution to the Company's earnings could be expected, a significant improvement in price would be required as well as the resolution of startup problems.

In the middle and late Sixties and early Seventies, the nation's petrochemical industry was experiencing a dramatic step-up. The market growth was accelerated by lower and lower prices for petrochemical products, making petrochemicals competitive in markets where they had not been before. This situation encouraged more and larger plants to be built; it seemed that everybody was getting into the act. As a result, output of the new plants and expanded capacities began to exceed the market demand.

Meanwhile, El Paso Products Company concentrated on developing a strong management team, de-bottlenecking and expanding its existing plants in Odessa. In some cases the existing plants were expanded to more than twice their original capacities. Also, many engineering and technical skills were applied to make the plants more efficient and competitive. It turned out to be a rewarding approach. The Products Company was able to pay off all its indebtedness as well as paying substantial dividends to the parent company.

During this same period, the joint venture with Dart Industries—in polyethylene and polypropylene—was expanding, selling its technical assistance and licensing its patents to major companies in Japan, Italy, France, Norway, as well as to other companies in the United States.

In addition, under the strong leadership of W. D. Noel, the Products Company and its joint venture partners include facilities to produce butadiene, styrene, ammonia, ethylene, propylene, low density polyethylene, polypropylene, adipic acid, hexamethylenediamine, and nylon polymer and fibers.

In 1976, the Dart-El Paso Products Company joint venture completed construction on a polypropylene plant which more

LAKESHORE COPPER PROPERTIES are jointly owned by
El Paso and Hecla Mining Company near Casa Grande, Arizona.

POLYPROPYLENE PLANT,
near Houston, was completed
in 1976 as a joint venture
of El Paso Products Company
and Dart Industries, Inc.

than doubled the joint venture's polypropylene capacity, at a new site at Bayport, Texas on the Houston ship channel. Construction also began on a new low-density polyethylene plant with Dart Industries at the same location, and production was scheduled to begin in 1978. The availability of feedstocks from nearby petrochemical plants and refineries as well as deep water transportation facilities, influenced the selection of this site for both of these new plants.

An expansion of the styrene facility at Odessa was scheduled for completion in 1978.

Consolidated Thermoplastics Company, owned jointly with Dart Industries, is a major producer of polyethylene film. A plant in Harrington, Delaware, opened in 1974 to serve eastern and southeastern markets, has been expanded more than two-fold to supply the growing needs of customers.

However, it is a brutal fact of life that not all ventures are destined to prosper and grow. The nationwide recession of 1974-75 affected all elements of the fiber and textile industries, including Beaunit Corporation, the Company's subsidiary engaged in the manufacture of synthetic fibers and textiles. Having already undergone hard times for a number of years, Beaunit was severely affected by a sharp drop in the demand for fibers and textiles in the latter part of 1974. Production at most of its textile plants and all fiber plants was curtailed throughout the first part of 1975 in order to bring inventories in line with market demand. But by

CONSTRUCTION proceeded in 1976 at liquefactio

cilities to be owned and operated by Sonatrach near Arzew on the Mediterranean coast.

1976 the continuing losses suffered by Beaunit, combined with the still-depressed state of the textile and fibers industry nationally, led to El Paso's decision to dispose of the company. The sale, to an affiliate of T. A. Associates, was made in 1977 with several members of Beaunit's top management becoming members of the purchasing group.

With the nation in the midst of a critical energy shortage in 1977, the future of the company's primary business — energy — would seem to pose a serious threat to its economic prosperity. But quite the contrary is true. The reason: In its gas operations, El Paso is considered a utility, and utilities, unlike manufacturing companies don't make money by buying gas cheap, selling it high, and possibly suffering from an inability to obtain the merchandise which they market. The name of the game is "rate base," that is, the capital a utility has invested in order to serve the public.

Whatever the price the Company pays for gas, that price—no more and no less—is passed on to the consumer. Consequently, El Paso neither profits nor loses by the price it pays for gas. The price of the gas is fixed by the FPC on the amount of capital the Company has invested in its facilities. The greater the rate base, the greater the opportunity for profit.

"So," according to Howard Boyd, El Paso's Chairman, "the developing energy crisis and the shortage of gas in the United States—rather than being a threat to the future prosperity of the Company, is, in fact, an opportunity the likes of which the Company never before enjoyed."

In past years, when El Paso Natural was expanding its system into California and other markets, the costs of the pipeline projects were considered huge. El Paso might add a new pipeline to California at a cost of $200 million, or even $300 million. Now, in order to deal with the declining gas reserves and meet the demands for gas, El Paso is looking at projects which run into the billions of dollars, and it has opportunities to add rate base in a magnitude that had never even been conceived.

In addition, El Paso is exploring a project jointly with the Japanese and the Russians. Under this plan, gas would be produced in the northern regions of eastern Siberia, transported by pipeline to a port on the East Coast of Russia, north of Vladivostock. At this point, the gas would be liquefied and half of it made available to ships supplied by Japan, and the other half would go to El Paso in ships that it would supply, to bring gas to the West Coast of the United States. The Company is also trying to secure natural gas in other parts of the world, including Iran. These projects are of such magnitude that they boggle the imagination and will require capital investments the like of which are unknown in the utility industry.

Under another plan, called the Trans-Alaska Gas Project, up to three billion cubic feet of gas a day from Alaska's North Slope oil fields would be made available to markets in Alaska and the lower forty-eight states. In 1977, El Paso's proposal and two competing proposals were before the FPC for approval.

With supplies of natural gas dropping seriously in the United States, El Paso is facing idle capacity in some of its existing pipelines which run from West Texas to California. So El Paso Natural Gas Company has entered into an agreement with The Standard Oil Company (Ohio) under which a portion of its pipeline facilities would be utilized as part of a project to transport Alaskan crude oil from a port in California to a terminal near Midland, Texas. From that point the oil could be moved through established oil pipeline networks to refinery complexes located in the Midwest, Gulf Coast and East.

"We feel," says Boyd, "that this project serves everyone's interest and will make possible the movement of that Alaskan oil at a much earlier date than would be possible if a new pipeline had to be built."

Knowledgeable people in government and industry are quick to recognize that there is no substitute for the natural gas that supplies forty million homes housing two-thirds of the population in the United States. But as the nation's gas shortage hit, El Paso was the first to apply to the FPC for authorization to build a coal gasification plant. Some years ago, the Company acquired a coal lease on the Navajo Indian Reservation in northwestern New Mexico, and tests proved up hundreds of millions of tons of coal. The vast coal deposits lie practically "on top" of El Paso's existing pipelines — which were beginning to suffer from idle capacity. It seemed natural that gas made from coal on the location could easily be transported to market, and a major

Travis Petty

IN 1976, Howard Boyd was bestowed with the Cross of the Knight of the Legion of
Honor of the French Republic. Because of the long-standing friendship between
France and the United States dating back to the American Revolution, a number of
Americans have been honored by the Order. General John Pershing and President
Dwight Eisenhower were among them. This time it was an American businessman.
Above, Mr. and Mrs. Boyd are shown with General de Boissieu who presented
the award at the palace in Paris where Napoleon inaugurated the Order.

program was begun to put together a gasification project. It was originally designed to make about 250 million cubic feet of gas a day, using the Lurgi process designed by a German engineering firm. Cost of the plant, when first proposed in 1971, was estimated to be about $250 million. Then it began to escalate indescribably. In the long course of delays brought about by environmental groups, the FPC, the Interior Department and the Navajo Tribe, those costs have skyrocketed to over a billion dollars. Boyd says, "The matter is still presently unresolved and I can say that all proposed coal plants are suffering for one reason or another. I remember a time, shortly after I came to work with El Paso in the early Fifties, that the Company was already exploring the feasibility of converting coal to natural gas. I told Mr. Kayser that the whole concept surprised me and wondered how it could ever be justified from an economic standpoint.

"Mr. Kayser's response was a reflection of his great foresight. He told me that gas was a wasting asset and that in his opinion there would be, in time, a shortage of natural gas in the United States. Keep in mind, this was somewhere around 1953 or 1954. I would be less than frank if I didn't confess that I had some skepticism about Mr. Kayser's view. At that time there was an oversupply of gas, and huge quantities of it were still being flared or vented to the air since there was no useful purpose to which it could be put. The feasibility studies continued until we were satisfied that it was technically possible to convert coal to gas —and then the knowledge was put on the shelf."

So, until some way is found to resolve the tremendous costs of coal plants, the difficulties of financing them in a tight money situation and the attitudes of government and environmental circles, no one can say what the future holds for coal gasification.

Carlton Homan, El Paso's Vice President who is in charge of engineering projects, feels strongly that coal will play an important part in the Company's future, in spite of what, in the mid-Seventies, seems like insurmountable odds. "In some ways," says Homan, "this energy crisis has been the best thing that ever happened to us.

"I don't know if we'll ever build a plant, but we've got plenty of coal. Sure, there are some hangups in mining coal, environmental hangups, but those can be ironed out with some compromises. The environment has to be kept clean. I can see a tremendous growth in the acquisition of coal because it makes up most of the energy resources in this country. If we have to, we can mine it and ship it by train to power plants, wherever they may be. One of these days, nuclear energy will come into its own, although it will probably be after our lifetime. It will be the predominant source of energy for the whole world. Another thing, Mr. Kayser told me back in '55, that one day each home will be powered by solar energy cells of some sort or another. And this could well be. He was always thinking that way. Actually, it's only today's economics that keeps us from utilizing solar energy."

In the immediate future El Paso's optimism is well-grounded in reality. Take, for example, the first Algerian LNG project alone: Assuming the full fleet of nine ships to be in operation at 100 percent load factor, the terms of the contract will generate $65 million a year, net after taxes. This is equivalent to, and even exceeds the Company's entire annual earnings in many past years. And this is just one of the behemoths that El Paso has in various stages of development.

In 1974, El Paso and its subsidiaries achieved the highest revenues and earnings in the history of the Company notwithstanding the divestiture of the Northwest Division pipeline system. In May, 1976 Chairman Howard Boyd announced the election of H.F. Steen to Vice Chairman of the Board of El Paso Natural Gas, and the elevation of Travis Petty to President and Chief Operating Officer of El Paso Natural. Steen would continue as President and Chief Executive Officer of the parent El Paso Company.

So the declining gas supply situation in the United States and the critical energy shortage provide opportunity for prosperity of unprecedented dimension. El Paso can very likely enjoy a future which had never even been dreamed of as recently as 1970.

EPILOGUE

Histories are not written about the present. But the present stands on the shoulders of the past. We look back over the decades and realize that what happened then has become history. At the time, it was merely a way of life. And in the case of El Paso Natural Gas it contained all the elements of a bigger-than-life drama. It runs the whole scale and it ranges from pioneering days in the oil fields, the bittersweet humor of men who fought the elements and the Depression, successes and failures, good times and bad.

And, as it should, El Paso Natural is changing. The men and women who put together its first, shaky foundation and made it into a great corporation are beginning to make way for a new generation to carry on in the same tradition. Paul Kayser, the unusual man who started it all, began this tradition in 1928 with a personal philosophy based on integrity, foresight and perseverance to pioneer the unknown. His first consideration, however, was that of the well-being of the people he gathered around him. Through the years, that consideration produced a sense of dedication and loyalty among the thousands of people who followed and that few, if any, organizations have ever enjoyed. These people realized that El Paso was *their* company. The words "labor" and "management" were seldom heard. The tendency was to work with somebody, rather than for somebody. These people worked long, hard hours—around the clock if necessary —to get a job done. And they did it because they wanted to.

Times change, people change. The original concept of El Paso Natural as a straight pipeline company is changing more rapidly today than ever before. What began as a "man's" company is now being shared by hundreds of women in the fast-paced computer age of finding ways to alleviate the nation's energy shortage in bold and imaginative ways.

But El Paso's special breed of people cling to the original values of giving their best as they probe new dimensions that no man has yet explored.

ACKNOWLEDGMENTS

Perhaps it is unique for an author of a book to incorporate another person's acknowledgments. However, this is a special opportunity not only to show my appreciation for those who have assisted with the gathering of facts and compilation of this material, but also to include comments of one of the earliest employees of the company, Hugh F. Steen, who became President as well as one of its "super-spokesmen". It is with great pleasure that I am able to include some of Steen's capsule thoughts covering his association with El Paso Natural Gas.

Steen says, "So many people have been helpful and taught so much (though I might unintentionally overlook many who made deep impressions), credit should be given to a few whose examples, advice, reprimands and counsel made it possible for me to thoroughly enjoy my years of service with EPNG:

Paul Kayser — whose character, examples and over-all ability made him a great leader.

Cyrus Perkins — whose ruggedness, innovative determination and character had a great deal to do with the growth and success of the company. A whole book could be written about his contribution, in attitude and leadership, to the building of the California system in record time.

Clarence Byrne — a diamond in the rough.

A. L. Forbes — quite a guy, you knew where you stood. He advised me to keep working and not to return to school. I appreciate that.

R. W. Harris — a great pipeliner and a very dedicated person. His sudden death was a terrible shock to all of us.

Walt Miller — they broke the mold. He was a great one and did a great job.

Other special people — C. C. Cragin, Catherine Martch, Max Norwood, Fred Wagner, John Eichelmann, Johnny Schaffer, J. E. Franey, Jack Stricklin and his dad, E. M.; Estin Scearce, Clarence Selley, Bob Cowan, D. H. Tucker, Marshall Willis, Norris Stogner, Ewell and Ray Walsh, Everett Woody, Virgil and Preach Rittmann, Johnny Crutcher, Jack Purvis . . . and hundreds of wonderful people in the Operating, Engineering and Accounting Departments. Most of these people were friends of mine and great people to know and have around. All of these and many more had a profound and lasting effect on my life and were to a great degree responsible for whatever small degree of success has come my way.

Burney Warren, Travis Petty, Bill Holik, Dick McConn, Homer Wilson, A. M. Derrick — along with their managers and superintendents, among whom are T. J. Crutchfield, Johnny Doan, Bill Parrish, and a host of others hard at work seeing that El Paso Natural continues to grow and prosper.

W. D. "Bill" Noel — his talented group has nurtured the Products Company from a modest inception to one of the principal components of the organization.

Howard Boyd — our chairman and chief policy officer keeps close tabs on everything that is going on in the organization. He and George Carameros along with Barry Hunsaker, and the outstanding team they have assembled, are in the process of bringing the LNG Company into fruition. This will be the largest project ever attempted by the Company and a history of the efforts expended here will someday be written. This will be a most interesting story.

As time passes, I am sure that I will remember the names of many people who influenced and helped me in many ways. Most of all I want to mention:

Mary Marshall — whom I married in 1937. She has been a true pipeliner in every respect, going along with the 'Company-comes-first' attitude — (at least most of the time)."

Putting the story of El Paso Natural Gas Company into its proper perspective was possible primarily because many of the original employees are living and well, and their memories — in some cases with near total recall — proved invaluable. In instances when versions conflicted slightly in detail, or memory was dimmed by time, I verified facts using corporate records, annual reports to stockholders, and copies of El Paso's employee magazine, *The Pipeliner*.

The cooperation and interest of many people from within and from outside the Company was extremely satisfying. It is not possible here to name them all; however, I wish to express a personal debt of gratitude to a special group of associates and friends for their help in producing this book:

To Hugh F. Steen who instigated the project.

To Mike Brumbelow who encouraged me to take on the job.

To John McFall for his confidence, encouragement, and understanding that publishing this book would take much of my time for several years.

To Jonell Haley and Kathy Edwards who collaborated with me closely on every aspect of this book—for typing and retyping the manuscript and many wise decisions. For their unmatched ability to find errors that creep insidiously into writing and proofreading and for their personal enthusiasm, they have my total admiration.

To members of El Paso's Public Relations and Advertising group whose expertise aided considerably, particularly Ruth Willis, Linda Bransford, Mike Taylor and writers Jim Collins, Joe Arnett, and Doug Meed, who read the manuscript and added their thoughts.

To my friend Leon Metz, noted Southwestern writer and historian, who spent many hours editing the manuscript.

To photographers Cliff Trussell, Cletis Reaves, Betty Baker, Jo Ann Stearn and David Shindo whose pictures and darkroom work are unmatched.

To Board Chairman Howard Boyd, who read the manuscript — along with Paul Kayser, Hugh F. Steen, John Eichelmann, C. L. Perkins, Travis Petty, Burney Warren, William V. Holik, George Carameros, E. G. Najaiko and Carlton Homan.

To the many people who gave their time for interviews and to those who provided historical pictures.

And finally, to El Paso Natural's first employee, Miss A. C. Martch, whose vivid memories, storehouse of facts, and boundless interest helped to make producing this book a pleasant and rewarding experience.

INDEX

SOURCES

American Graphics Press: Dallas, 1975.

Arizona Republic.

Big Lake *Wildcat,* The.

Casa Grande Dispatch.

Cleland, Robert Glass, *A History of Phelps Dodge.* New York: Alfred A. Knopf, 1952.

El Paso *Herald Post.*

El Paso *Times.*

Facts and Fantasies of the Oil Patch. Oklahoma City: Desk and Derrick Club, 1975.

Forbes.

Garraty, John, *The American Nation.* New York: Harper and Row, 1966.

Gibson, Jewel, *Black Gold.* New York: Random House, 1950.

Griffin, John H., *Land of the High Sky.* First National Bank of Midland.

Handbook of Texas. Austin: The Texas State Historical Association, 1952.

Knowles, Ruth, *America's Oil Famine.* New York: Coward, McCann and Geoghegan, Inc., 1975.

Midland *Reporter Telegram,* The.

Myres, Samuel D., *The Permian Basin.* El Paso: Permian Press, 1973.

Odessa *American,* The.

Phoenix Gazette.

Pope, Clarence, *An Oil Scout in the Permian Basin.* El Paso: Permian Press, 1972.

Rister, Carl Coke, *Oil! Titan of the Southwest.* Norman: University of Oklahoma Press, 1949.

San Angelo Standard Times.

Standard Oiler, The. San Francisco: Standard Oil Company of California.

Texas Almanac.

Time.

Time-Life Books.

Tompkins, Walter A., *Little Giant of Signal Hill.* Englewood Cliffs, New Jersey: Prentice-Hall, Inc., 1964.

World Book Encyclopedia, The. Chicago: Field Enterprises Educational Corporation, 1968.

ABOUT THE PICTURES

The photographs in this book came from many sources. The oldest pictures were graciously loaned by various newspapers, particularly the *San Angelo* (Texas) *Standard Times*. Others were furnished by employees of El Paso Natural Gas Company who had the foresight to take snapshots of operations and installations as early as 1929. Even so, the editorial choice was slim.

In the early 1950s, El Paso Natural formed its own photographic section, headed by Cliff Trussell and Cletis Reaves, two of the Southwest's finest photographers. They continue to record on film, every major project of The El Paso Company and its affiliates, and consequently, thousands of first-rate photographs were available from their files for use in this history.

This Book

is lithographed by Guynes Printing Company, El Paso, Texas.
The text is set in 11-point Baskerville, leaded two points.
Paper is 80-pound Quintessence Dull, by Northwest.
Binding is by Gerhard Schermer.